GIScience and Geo-environmental Modelling

The "GIScience and Geo-environmental Modelling" book series seeks to publish a broad portfolio of scientific books addressing the interface between geography and the environment. The aim of the book series is to present geospatial technology approaches to data mining techniques, data analytics, modeling, risk assessment, visualization, and management strategies. The series includes peer-reviewed monographs, edited volumes, textbooks, and conference proceedings. The focus of Geo-environmental is on geospatial modelling in the frontier area of earth-environment related fields, such as urban and peri-urban environmental issues, ecology, hydrology, basin management, geohazards, estuarine-ecology, groundwater, agriculture, climate change, land-water, and forest resources, and related topics.

Geo-environmental modelling techniques have enjoyed an overwhelming interest in recent decades among the earth environmental and social sciences research communities for their powerful ability to solve and understand various complex problems and develop novel approaches toward sustainable earth and human society. Geo-environmental modelling using data mining, machine learning, and cloud computing technology is focused on spatiotemporal data analysis and modeling for sustainability in our environment.

Swades Pal · Swapan Talukdar ·
Tamal Kanti Saha · Rajesh Sarda

Damming and Wetland Eco-hydrology

Evaluation, Resilience and Restoration

Springer

Swades Pal
Department of Geography
University of Gour Banga
Malda, West Bengal, India

Swapan Talukdar
Department of Geography
Ashutosh College
Kolkata, West Bengal, India

Tamal Kanti Saha
Department of Geography, Faculty of Sciences
Jamia Millia Islamia
New Delhi, Delhi, India

Rajesh Sarda
Department of Geography
University of Delhi
New Delhi, Delhi, India

ISSN 2730-7506 ISSN 2730-7514 (electronic)
GIScience and Geo-environmental Modelling
ISBN 978-3-032-11728-1 ISBN 978-3-032-11729-8 (eBook)
https://doi.org/10.1007/978-3-032-11729-8

This Springer imprint is published by the registered company Springer Nature Switzerland AG
The registered company address is: Gewerbestrasse 11, 6330 Cham, Switzerland

To our families, whose unwavering support made this possible.

Foreword

It is a privilege to introduce the book volume *Damming and Wetland Eco-hydrology: Evaluation, Resilience and Restoration*, a timely and important contribution to the growing body of literature at the intersection of hydrology, ecology, and environmental management. This work, authored by Prof. Swades Pal, Dr. Swapan Talukdar, Dr. Tamal Kanti Saha, and Mr. Rajesh Sarda, presents a comprehensive and methodologically robust exploration of the multifaceted impacts of damming on wetland systems, with a specific focus on riparian and floodplain environments.

The book offers an integrated assessment of hydrological alterations caused by dam structures and their cascading effects on wetland ecology. With an emphasis on channel flow dynamics, wetland habitat transformation, eco-hydrological modeling, and resilience strategies, the authors have successfully blended empirical case studies with advanced analytical tools, including the application of artificial intelligence and remote sensing technologies. What distinguishes this volume is its interdisciplinary approach and its commitment to both scientific depth and applied relevance. The book addresses key contemporary challenges, including water scarcity, wetland degradation, and the ecological consequences of disrupted river-wetland connectivity. This book not only documents change but also proposes innovative frameworks for eco-flow assessment and restoration grounded in nature-based solutions.

This book will serve as a valuable resource for graduate and postgraduate students, researchers, environmental professionals, and policymakers engaged in water resources management, wetland conservation, and sustainable development. It exemplifies academic scholarship that is both theoretically informed and practically oriented, making a meaningful contribution to the discourse on ecological sustainability in the Anthropocene.

I congratulate the ***editors*** for their foresighted effort in bringing out this book volume at a very right time on such a pertinent and burning issue of wetland ecosystem and eco-hydrology. My best wishes for their future academic endeavors and wish that this book will be very useful to the scientific communities and scholars across the age and discipline.

Prof. Atiqur Rahman
Head and Professor, Department
of Geography, Faculty of Sciences
Jamia Millia Islamia
New Delhi, India

Preface

The book on *Damming and Wetland Eco-hydrology—Evaluation, Resilience and Restoration* explores the complex impact of damming on river and wetland eco-hydrology considering different facets. In order to meet the growing demand of water and energy footprint, among different adopted measures, damming over river is one of the significant measures all across the world. Most of the large rivers, even small rivers, witnessed huge hydrological modification through damming. The major reasons behind damming are enhancing water security in different required sectors and energy generation. Such measure has offered a wide range of benefit to the human society along with a number of burdens to the river and dependent wetland hydrological and ecological conditions. At first the book examines the impact of damming on flow dynamics and eco-hydrological ambience of river and the later part explores the effect of river water change on wetland eco-hydrological situation. Since wider number of marginal people including fishermen, farmer communities all across the world strongly depend on wetland for their livelihood and nutrition, loss, degradation of wetland health may cause crisis in their lives. So, scientific assessment of multifacet impacts of hydrological modification in dam downstream river is very vital for adopting suitable measures minimizing and mitigating the emerging or emerged crises. It is essential for the long-term sustainability of the riparian natural resources and communities.

Key features of this book with nine chapters are (1) de-knot complex relation between dam, river, and wetland focusing both hydrological and ecological aspects, (2) wide range of scientific, robust methods adopted for exploring the objectives, (3) pixel-scale hydrological, eco-hydrological modeling of wetland, (4) python language-driven parameter-based wetland health modeling highlighting different hydrological, morphological, ecological, and pollution components, (5) paradigm shift of modeling approach with applications, (6) it is a good blend of topical and methodological aspects on the topic.

This book would be a good guide to the scholars wishes to work on this issue and related geo-environmental issues. Since the book carefully highlighted language-based modeling approach for wetland health modeling, this script would be beneficial not only to the scholars who are working in this specific field, but also it would be valuable to those who are even working on modeling environmental hazards, risks, vulnerability, resilience, etc.

Malda, India — Swades Pal
Kolkata, India — Swapan Talukdar
New Delhi, India — Tamal Kanti Saha
New Delhi, India — Rajesh Sarda

Acknowledgements Writing a book is really harder than we thought and rewarding than we could have ever imagined. None of these would have been possible without a credential academic bonding and knowledge support among authors. First of all we are deeply grateful to Doris Bleier, Publishing editor and Biswajeet Pradhan, Pravat Kumar Shit, Gouri Sankar Bhunia, Partha Pratim Adhikary, Hamid Reza Pourghasemi, Series Editor of the book series entitled *GIScience and Geo-environmental Modelling* for giving us opportunity to work on this book and extending invaluable guidance and support throughout the project. We also want to extend our heartfelt gratitude to our family members for their unwavering support and encouragement. Without their supports, it would be very difficult to devote ourselves on this work and make it successful. We are indebted to Ripan Ghosh, Sudarshan Sarkar, scholar, and student of the University of Gour Banga for their unconditional assistance to develop some of the maps in this project. We must thank our institutions for promoting us for involving with such a noble work.

Competing Interests The authors have no competing interests to declare that are relevant to the content of this manuscript.

Contents

About the Authors

Swades Pal is a professor in the Department of Geography, University of Gour Banga, India. His research interest is on river and riparian wetland hydrology and ecology. He tries to explore the multi-dimensional impacts of climate change, human interventions in floodplain wetland ecosystem, and socio-ecological system. His name was listed in the top 2% of world scientists (2020, 2021, 2022, 2023, and 2024) published by Stanford University, USA. Further information about his work and previous publications can be found here: https://scholar.google.com/citations?user=XvRpYXEAAAAJ&hl=en

Dr. Swapan Talukdar is an assistant professor in the Department of Geography at Asutosh College, University of Calcutta, Kolkata, India. He has successfully completed two international projects (as Co-Principal Investigator) in collaboration with King Khalid University, Saudi Arabia. His primary research interests include urban environments, urban wetlands, natural hazards, remote sensing, and artificial intelligence applications in environmental management. Further information about his work and previous publications can be found here: https://scholar.google.co.in/citations?user=e9MgpZkAAAAJ&hl=en

Dr. Tamal Kanti Saha is a SERB-DST National Post Doctoral Fellowship (N-PDF), currently affiliated with the Department of Geography, Faculty of Science, Jamia Millia Islamia, New Delhi. His research interest is on wetland hydrology and ecology. He involves exploring regional-level drivers and dimensions of wetland transformation. Further information about his work and previous publications can be found here: https://scholar.google.co.in/citations?user=KmyUEp8AAAAJ&hl=en

Rajesh Sarda is currently pursuing Ph.D. degree from the Department of Geography, Delhi School of Economics, University of Delhi, Delhi, India. He completed his M.Phil. degree from the Department of Geography, University of Gour Banga, West Bengal, India. His research interests focus on wetland hydrology, wetland ecosystem service, natural hazards, and application of AI for environment management. Further information about his work and previous publications can be found here: https://scholar.google.com/citations?user=7j0CJZcAAAAJ&hl=en

List of Figures

List of Tables

Understanding Interplay Among Dam, River, and Wetland

1

Abstract

This chapter has examined how damming influences on river flow and, in turn, how these affect hydrology and health of floodplain wetland. Review works indicate that damming on river has growingly been appearing as a major global concern. Theoretical interplay among dam-river-wetland has been explored through some specific mechanisms like seepage, flooding, groundwater support, tie-channel connection, etc. In order to explore this relation, wetland-dominated Atreyee River basin spanning between India and Bangladesh was selected, as it underwent damming in 2012, leading to significant hydrological medications in river and riparian systems.

Keywords

Damming effect · River flow change · Floodplain wetland modification · Dam-river-wetland trinity

1.1 Concept of Damming

According to *Britannica*, a dam is a structure built across a river to hold back water and form a reservoir. Dams may range in size from very large to very small, and they are generally constructed to support irrigation, supply domestic water, and generate hydro-electricity. As per the International Commission on Large Dams (ICOLD 2021), a large dam is defined as one with a height of 15 m or more from the foundation. It may also include dams with a height between 5 and 15 m if they impound more than 3 million cubic meters of water. The practice of dam construction dates back to around 3000 BC during the ancient Mesopotamian civilization (Baba et al. 2018). The oldest known dam is the Jawa Dam, built across Wadi Rajil in Jordan (Fahlbusch 2009). Over time, with the progress of human civilization, the number of dams has multiplied many times. ICOLD identifies various types of dams, including earth, rock-fill, gravity, buttress, barrage, arch, and multiple arch structures. These dams may be constructed for a single purpose or for multiple purposes. In the last three centuries, freshwater extraction has increased by a factor of 35, while the world population has increased only by a factor of 8 (Mishra 2023). This growing demand has compelled societies to build numerous dams of different sizes across rivers. At present, ICOLD records account for 61,988 dams worldwide. Their spatial distribution is strongly correlated with national gross domestic product (GDP) (Fig. 1.1). Figure 1.2 further demonstrates the global damming scenario in relation to GDP. The usefulness of dams in today's world is well recognized. They provide essential services for water supply, power generation, and irrigation. However, their adverse impacts—such as altering river flow distribution, causing geomorphic changes, and affecting floodplain resources—cannot be ignored.

S. Pal et al., *Damming and Wetland Eco-hydrology*, GIScience and Geo-environmental Modelling,
https://doi.org/10.1007/978-3-032-11729-8_1

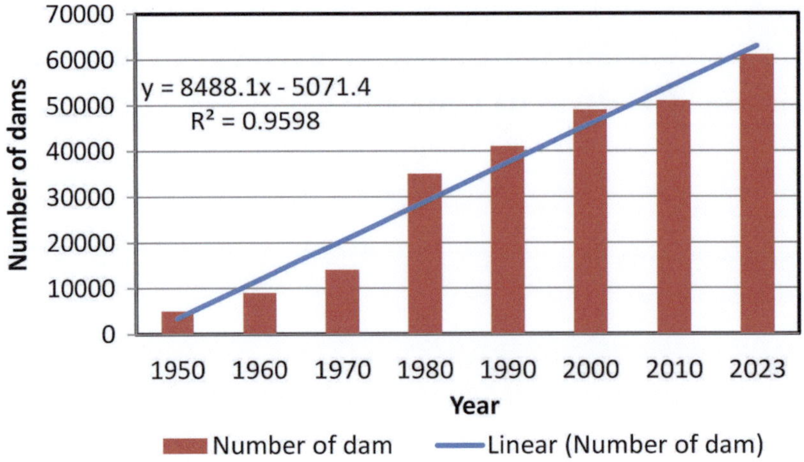

Fig. 1.1 Changes in dam frequency of the world from 1950 to 2023

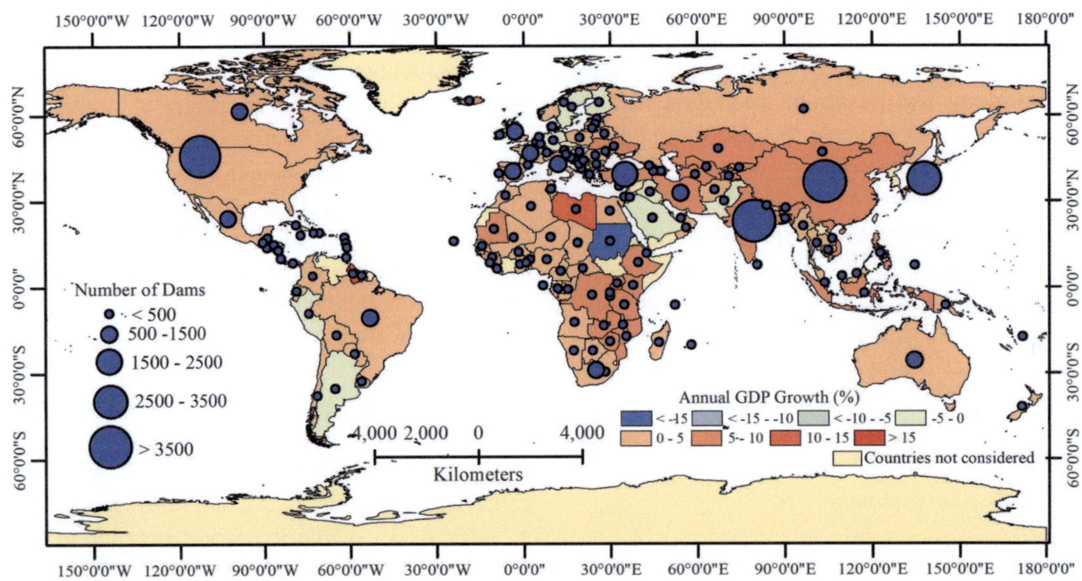

Fig. 1.2 Country-scale scenario of damming frequency in reference to growth rate of gross domestic product (GDP)

1.2 Benefits and Burden of Damming

Damming programs were initiated and continued across the world to provide a wide array of socio-ecological benefits. These include flood control, regulated flow allocation, hydro-electricity generation, increased freshwater supply for agriculture, industry, and domestic use, and overall enhancement of water security (Everard 2019; Vidal-Abarca Gutiérrez et al. 2023). Such benefits have made dams central to modern water management and development strategies.

However, alongside these advantages, damming has also transmitted several burdens to geomorphic, hydrological, and socio-ecological systems. These include fragmentation of river habitats due to longitudinal disconnection (Barbarossa et al. 2020), reduction in sediment supply affecting downstream ecosystems (Larsen et al. 2021), and alteration of natural flows (Lai et al. 2021). Dams also modify inundation regimes (Jiang et al. 2023), cause reservoir backwater storage (Hunt et al. 2020), induce riverbed aggradation (Liu et al. 2023), and result in changes to the wetted perimeter and irregularities in flow magnitude, duration, and timing (Almeida et al. 2020) (Table 1.1). These physical and hydrological disruptions lead to further consequences, including negative impacts on aquatic reproduction and migration, soil erosion, altered geochemistry of water, changes in physical composition, modifications to the micro-thermal environment, lowering of groundwater levels, habitat disturbance, and biodiversity loss. In addition to affecting downstream river courses, these impacts extend to floodplain resources, ecosystems, and dependent human communities. Dam-driven displacement of population remains another critical social issue that has rarely been resolved successfully.

1.3 Wetland: Concept, Importance, and Loss Scenario

A considerable number of definitions have been given by scholars and reputed national and international institutions. In simple terms, wetlands are described as the intermediate world or ecotone between land and water ecosystems, existing along a continuum ranging from permanent wetlands to those that inundate only once in a year (Mitsch and Gosselink 2007). The Ramsar Convention (1971) defined wetland as "areas of marsh, fen, peatland or water, whether natural or artificial, permanent or temporary, with water that is static or flowing, fresh, brackish, or saltwater, including areas of marine water, the depth of which at low tide does not exceed six meters." Thus, wetlands can be understood as areas that remain flooded or saturated by surface water, groundwater, or seawater at different frequencies and durations, and that share characteristics of both terrestrial and aquatic ecosystems. The occurrence of water, the presence of hydrophytes, and hydric soil are considered the three basic criteria for identifying wetlands. Different categories include swamps, marshes, peatlands, fens, and bogs (Dungan 1990).

Wetlands are recognized as important habitats because their unique hydrology and saturated soils support a wide variety of biodiversity (Mitsch and Gosselink 2000). Floodplain wetlands, in particular, hold immense ecological and social significance (Li et al. 2015; Mondal and Pal 2017). The well-being of the human ecosystem is closely linked with this natural capital, as wetlands provide a wide range of ecosystem services. These include the supply of food resources, purifica tion and regulation of water, refinement of air, carbon sequestration, climate regulation, nutrient cycling, flood control, and biodiversity management (Mitsch et al. 2010; Knight et al. 2017).

However, the perception of wetlands in society is often shaped only by their tangible economic returns, which are not always immediate or high. This misjudgment has led to their neglect. The Millennium Ecosystem Assessment (MEA 2005) report highlighted that beyond the intangible benefits, wetlands can yield 2.5 times higher economic revenue than agricultural land of the same unit area because of their ability to provide a wide range of goods. In this sense, wetlands are irreplaceable ecosystems, as their ecological and hydrological functions cannot be fully substituted by any other form of ecosystem (Mitsch and Gosselink 2000).

The benefits provided by wetlands are generally termed ecosystem services (United States Environmental Protection Agency 2009). These are usually grouped into four categories: provisioning, regulating, habitat, and cultural services. Table 1.2 illustrates the specific services provided by wetlands. The estimated global value of these services is 47.4 trillion US dollars per year, which accounts for about 43.5% of the total value of all biomes (Davidson et al. 2019). Table 1.3 shows the comparative economic values of different types of wetlands. Despite covering only 6% of the Earth's surface (Mitsch et al. 2010), wetlands contribute nearly 40% of the ecosystem services globally (Zedler and Kercher 2005).

India, with its diverse landscapes and climatic systems, sustains a wide variety of wetland habitats (Prasad et al. 2002). These wetlands support enormous biodiversity (Space Applications Centre 2011) and are therefore often

Table 1.1 Some major explored benefits and burdens of damming

Impact on	Specific impact	References
Channel morphology	Channel narrowing	Church (1995)
	Reduction of wetted channel width	Dey et al. (2020)
	Adjusted channel profile	Thoms and Walker (1993)
	Tributary incision due to base level change of the main river	Germanovski and Ritter (1988)
	River bed aggradations	Fergus (1997)
	Reduction in channel mobility	Gurnell et al. (1994)
	Narrower equilibrium channel formation	Lyons (1992)
	Back water stagnation upstream of reservoir	Graf (2006)
	Formation of artificial base level of erosion for the dam upstream reach	Ma et al. (2024)
Channel flow	Reduction in flow volume and water level	Pal et al. (2016)
	Reduction in flow velocity	Kulcsár et al. (2012)
Flow regime	Adverse change in flow regime (magnitude, frequency, duration, rate)	Graf (2006)
	Reduction in flow variability	Graf (2006)
	Reduction of seasonal flow variability	Graf (2006)
	Increase of low-flow pulses	Graf (2006)
	Reduction of high-flow pulses	Graf (2006)
Chanel inundation	Flood reduction (magnitude, frequency, duration, extent)	Davies et al. (2000)
Sediment supply	Sediment trap in reservoir	Tan et al. (2019)
	Reduction in downstream sediment supply	Lu and Siew (2006)
In-channel habitat	Water stress in downstream reach	Munia et al. (2016)
	Longitudinal disconnection and hindrance against species migration	Fullerton et al. (2010)
Riparian zone	Reduction in regular water supply between river and floodplain	Opperman et al. (2009)
	Reduction in nutrient supply	Morris et al. (2013)
	Squeeze of water body area	Wu et al. (2022)
	Weakening of wetland health	Chen et al. (2019)
	Adverse change in eco-hydrological condition	Sarda and Pal (2022)
	Lateral disconnection between river and many water bodies	Brunner et al. (2009)
	Riparian vegetation loss	Guo et al. (2015)
	Increasing stress on floodplain habitat	Suchara (2019)
	Shrunken riparian habitat	Graf (2006)
	Simplified array of habitat patches	Graf (2006)
Ecosystem services	Reduction of provisioning services (fishes/vegetables)	Maithya et al. (2022)
	Reduction of habitat services	Rahimi et al. (2020)
	Reduction of cultural services	Clarke et al. (2021)

Table 1.2 List of ecological integrity and ecosystem service components with rationale

Ecosystem service group	Service	Justification	References
Regulating services	Global climate regulation	Wetlands help mitigate climate change by sequestering carbon and reducing greenhouse gas emissions	Burkhard et al. (2012), Montoya-Tangarife et al. (2017)
	Local climate regulation	Vegetation in wetlands cools local temperatures by shading and evapotranspiration	Vihervaara et al. (2010), Burkhard et al. (2012)
	Air quality regulation	Wetland plants purify air by absorbing pollutants and releasing clean oxygen	Burkhard et al. (2014), Lin et al. (2019)
	Water flow regulation	Wetlands absorb runoff, reducing peak flood levels and controlling water flow	Burkhard et al. (2014), Montoya-Tangarife et al. (2017)
	Water purification	Wetlands naturally clean water, though they may sometimes release impurities	Burkhard et al. (2014), Montoya-Tangarife et al. (2017)
	Waste treatment	These systems break down and neutralize pollutants and excessive nutrients	Burkhard et al. (2014), Lin et al. (2019)
	Nutrient regulation	Wetlands support nutrient cycling, helping maintain ecological balance	Burkhard et al. (2012), Quyen et al. (2017)
	Natural hazard regulation	Wetlands buffer communities from floods by storing water and reducing flow energy	Quyen et al. (2017), Burkhard et al. (2014)
	Pollination	Wetlands sustain pollinators essential for agricultural productivity	Bartholomew et al. (2019), Lin et al. (2019)
	Pest and disease control	Diverse ecosystems suppress disease vectors like mosquitoes	Koschke et al. (2012), Quyen et al. (2017), McInnes and Everard (2017)
	Erosion prevention	Vegetation holds soil in place, reducing erosion from wind and water	Koschke et al. (2012), Olson et al. (2017)
Provisioning services	Food	Wetlands offer fish, crops, fruits, and other edibles	Koschke et al. (2012), Pritchard et al. (2019)
	Biomass for energy	Wetland plants can be used as renewable biofuel sources	Vihervaara et al. (2010), Koschke et al. (2012)
	Fodder	Provide grass and plant materials for animal feed	Burkhard et al. (2014), Lin et al. (2019)
	Livestock	Habitat and forage for raising domesticated animals	Burkhard et al. (2014), Montoya-Tangarife et al. (2017)
	Fiber	Supplies materials like wool and timber for construction or textiles	Olson et al. (2017), Pritchard et al. (2019)
	Timber	Trees and shrubs provide wood resources	Burkhard et al. (2014), Montoya-Tangarife et al. (2017)
	Wood fuel	Provides firewood, peat, and other biofuels	Pritchard et al. (2019), Li et al. (2018)
	Fish and edible algae	Aquatic species and algae serve as food sources	Wondie (2018), Li et al. (2018)
	Medicinal resources	Wetland plants are used in traditional and modern medicine	Burkhard et al. (2012), Pritchard et al. (2019)
	Freshwater supply	Used for drinking, irrigation, and livestock needs	Burkhard et al. (2012, 2014)
	Mineral resources	Extracted for economic or industrial purposes	Burkhard et al. (2014), Lin et al. (2019)

(continued)

Table 1.2 (continued)

Ecosystem service group	Service	Justification	References
	Transportation	Wetlands facilitate boat travel and water-based transport	Bartholomew et al. (2019), Lin et al. (2019)
	Abiotic energy sources	Energy harnessed from wind, water flow, or geothermal in wetland areas	Koschke et al. (2012), Quyen et al. (2017)
Cultural services	Recreation and tourism	Wetlands provide venues for leisure, such as boating or sightseeing	Wondie (2018), Lin et al. (2019)
	Aesthetic and inspiration	Scenic landscapes inspire art and contribute to mental well-being	Koschke et al. (2012), Quyen et al. (2017)
	Knowledge systems	Used as outdoor classrooms and research sites	Koschke et al. (2012), Kopperoinen et al. (2014)
	Religious and spiritual	Sacred places or traditional ceremonies tied to wetland locations	Koschke et al. (2012), Quyen et al. (2017)
	Cultural heritage	Preserve traditional knowledge and historical land uses	Burkhard et al. (2012), Montoya-Tangarife et al. (2017)
	Natural diversity	Home to a variety of flora and fauna, contributing to biodiversity	Koschke et al. (2012), Kopperoinen et al. (2014)
Habitat/ support services	Biodiversity conservation	Support diverse and sometimes rare species	Burkhard et al. (2012), Quyen et al. (2017)
	Nursery ground	Safe breeding grounds for fish, birds, and amphibians	Bartholomew et al. (2019), Lin et al. (2019)
	Food source for species	Interconnected species provide a robust food web	Burkhard et al. (2012), Montoya-Tangarife et al. (2017)
	Water quality improvement	Trap sediments and pollutants, improving overall water clarity	Lin et al. (2019), Burkhard et al. (2014)
	Flood control	Temporarily hold water, reducing downstream flood intensity	Quyen et al. (2017), Montoya-Tangarife et al. (2017)
	Shoreline stabilization	Vegetation protects coasts and riverbanks from erosion	Koschke et al. (2012), Olson et al. (2017)

described as "cradles of biodiversity," "biological supermarkets," "kidneys of the landscape," and "repositories of resources." Such metaphors reflect their ecological richness and functional indispensability.

In relation to dam–river–wetland interaction, multiple mechanisms are at play. Seepage from reservoirs or dam walls can recharge adjacent wetland areas, but excessive or poorly managed seepage may also cause waterlogging and salinity that degrade the wetland ecosystem. Flooding, controlled or accidental, often spreads dam water over floodplains and wetlands, which can replenish soil nutrients and maintain habitat diversity; however, altered flood timing may disturb natural breeding cycles of aquatic and terrestrial organisms. Groundwater support provided by regulated dam releases sustains wetlands during dry seasons, but long-term changes in recharge rates can lower groundwater levels and threaten wetland persistence. Finally, tie-channel disconnection, where natural connecting channels between rivers and wetlands are cut off by dam regulation, leads to isolation of wetlands. This isolation reduces the exchange of sediments, nutrients, and aquatic species, thereby weakening the resilience of wetland ecosystems. These mechanisms, shown in Fig. 1.3, demonstrate that the relationship between dams, rivers, and wetlands is complex and requires careful management to balance ecological and societal needs.

Loss of wetland and qualitative degradation of wetland scenarios are very unfortunate in spite of its immense ecological values. Fluet-

Table 1.3 Estimated ecosystem service value of different wetland types after (Davidson et al. 2019)

Wetland category (Davidson and Finlayson 2018)		Biome (Costanza et al. 2014)	Average 2011 unit value (2007 Int$ ha^{-1} yr^{-1})	Area ($\times 10^6$ ha)	Total global 2011 value ($\times 10^{12}$ Int$ yr^{-1})
Coastal wetlands	Coral reefs	Coral reefs	352,915	28.4	10.02
	Seagrass beds	Coastal systems*	28,917	78.8	2.28
	Estuaries	Coastal systems*	28,917	66	1.91
	Salt marshes	Coastal wetlands**	193,845	5.5	1.07
	Unvegetated tidal flats	Coastal wetlands**	193,845	12.8	2.48
	Mangroves	Coastal wetlands**	193,845	13.8	2.68
Total coastal wetlands				210.2	20.44
Inland wetlands	Tropical forested peatlands	Tropical forest	5382	36.2	1.95
	Temperate and boreal forested peatlands	Temperate and boreal forest	3137	33.4	1.05
	Non-forested peatlands	Swamps and floodplains	25,682	311.8	8.01
	Forested wetlands (Alluvial)	Tropical/temperate/ boreal forest	3460	117	4.05
	Marshes and swamps (Alluvial)	Swamps and floodplains	25,682	253	6.5
	Rivers and streams	Lakes and rivers	12,512	64.3	0.81
	Natural lakes	Lakes and rivers	12,512	371.6	4.65
Total inland wetlands				1187.3	27
Total wetlands				1397.5	47.44
Other natural biomes		Tropical dry forest	5382	1161	6.25
		Temperate and boreal dry forest	3137	2913	9.14
		Grasslands and rangelands	4166	4418	18.41
		Open ocean	660	33,200	21.91
		Continental shelf	2222	2660	5.91
Total other biomes				44,352	61.62
Grand total (wetlands + other biomes)				45,749.50	109.06

Source Davidson et al. (2019)
* Including seagrass meadows, shallow coastal seas over continental shelves, rocky shorelines, and sandy beaches (De Groot et al. 2012)
** Including intertidal zones, including tidal marshes and mangrove ecosystems (De Groot et al. 2012)

Chouinard et al. (2023) clearly noted that since 1700, 3.4 million km^2 (confidence interval 2.9–3.8) of inland wetlands have been lost. Since 1970 to present about 50% wetland has been lost. The rate of loss is not uniform across the world. Let and Pal (2023) reviewed literatures on this and reported that the rate of loss is highly related with socio-ecological scenario of the countries concerned (Fig. 1.3). As a vector of wetland loss, different causes were identified; however, reclamation of wetland through agricultural land is very dominant. It is also highly different across the world. From this, it is very clear that agricultural expansion, drainage modification are dominant

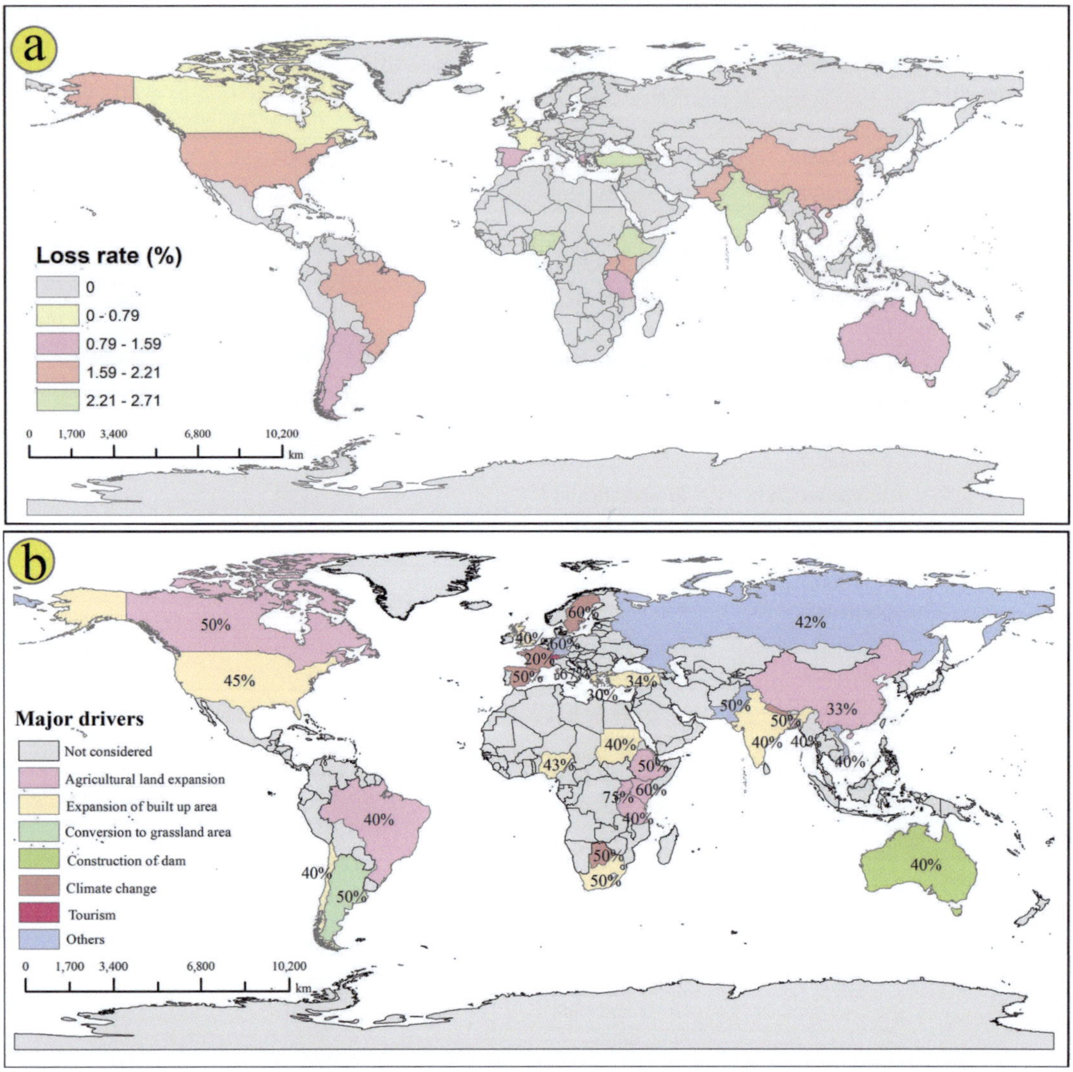

Fig. 1.3 Depicts **a** the yearly percentage of wetland degradation and **b** highlights key contributing factors. The percentages represent how frequently each driver was cited as the primary cause of wetland loss in various countries, based on literature analysis. Adapted from Let and Pal (2023)

causes of wetland conversion in developing and least developed countries. Contrastingly, pollution, built area expansions are found dominant reasons behind wetland transformation in developed countries (Fig. 1.3).

1.4 Dam, River, and Wetland Interplay

Dam restricts downstream natural flow and thereby controls in-channel hydraulics and out-channel flood regimes, flood extents. Floodplain is a low-lying area adjacent to river and regularly flooded during high discharge time. Wetland is one of the major components of floodplain. In fluvial region, particularly in the case of seasonal river, river is mainly fed by rainwater. Extreme

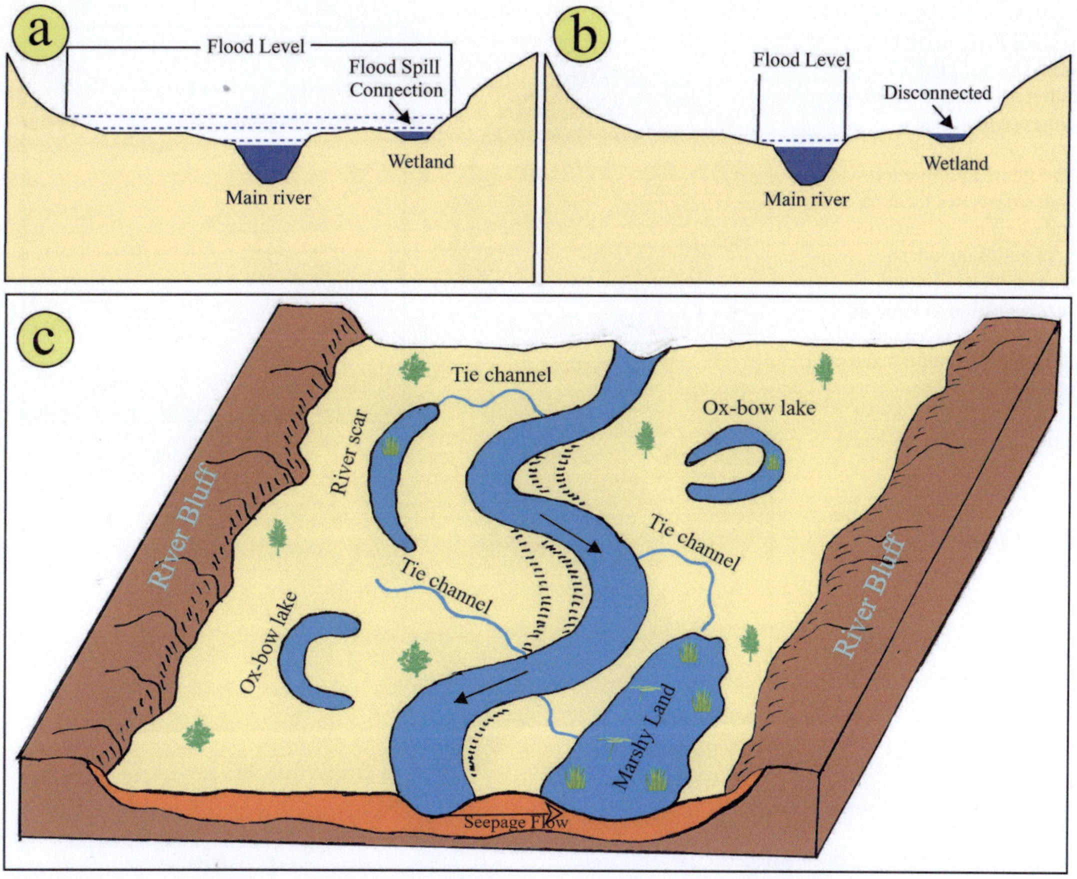

Fig. 1.4 Schematic diagrams showing floodplain hydrological components and damming effects **a** possible water level and flood extent in a floodplain in a unregulated river **b** possible water level and flood extent in a floodplain in a regulated river **c** schematic diagram of major hydrological components of floodplain and linkage between river and wetland

rainfall often creates high discharge which in turn spills to the floodplain based on the magnitude of spilling. Wetland in such river basin receives water from three major sources, including rainfall, discharge, and groundwater. So, the wetlands in the floodplain region connect themselves with river through flooding during flood time, river water seepage, and tie-channel flow. Tie channel is a connecting channel between river and wetland. It connects wetland with river hydrologically, sedimentologically and ecologically. During high discharge in river, a good amount of flow reaches to the wetland through tie channel and enriches wetland hydrologically (Let et al. 2024). Along with this sediment, different forms of fish pawns, others species and foods reach to the wetland. All these most of the cases enhance the sustainability of the wetland. Damming over river often causes lowering of discharge, water level, sediment flow downstream which lowers down the frequency of flooding, squeezes the lateral extent of flooding. As result of this, wetland outside the squeezed lateral flood limit suffers from water scarcity. Not only has that wetland within active flood-prone area also experienced huge hydrological irregularities. All these are against of natural hydrological and ecological regimes of the wetland. Figure 1.4 depicts the interplay between river and wetland in pre- and post-dam conditions.

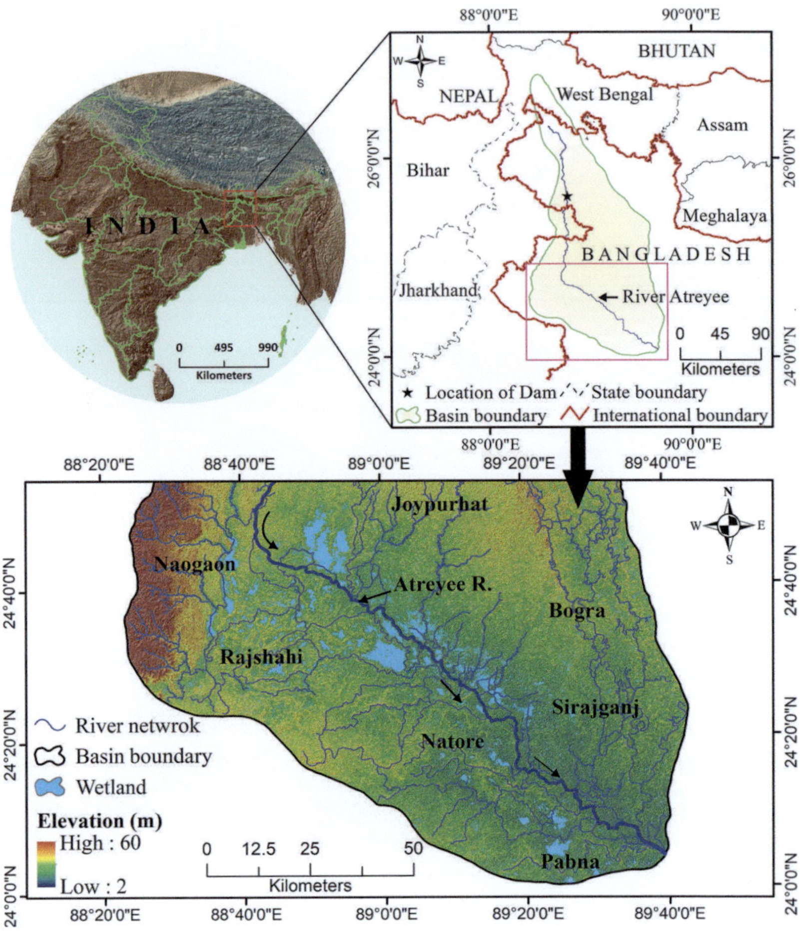

Fig. 1.5 Lower part of Atreyee River basin of India and Bangladesh taking as a case is shown in reference to broader geographical reference. First figure in upper left hand side shows location of study area in reference to India and Bangladesh, upper right hand figure shows position of Atreyee River basin in reference to India and Bangladesh and figure below shows wetland-dominated lower part of Atreyee River basin

1.5 Basic About Case Study Area

In order to explore the interplay among damming, river, and wetland, the Atreyee River basin of India and Bangladesh was taken into consideration. This river is an international watercourse extending approximately 390 km from Sevoke in West Bengal, India (26°48′46.14″, 88°29′13.68″E), to its confluence with the Brahmaputra River (locally known as the "Jamuna") near Ratanganj in Bangladesh (24°4′16.18″N, 89°39′34.60″E). Along its path, it traverses the Chalan Beel, a prominent sag pond in the Bangladeshi landscape (Rashid et al. 2014). The river basin spans between latitudes 24°1′07″N and 26°50′17″N and longitudes 88°22′45″E to 89°43′38″E, covering a total area of 22,054.47 km^2. It encompasses territories in four districts of West Bengal and fourteen districts within Bangladesh (see Fig. 1.5). A significant portion of its course lies within the Barind Tract, an ancient elevated alluvial region.

Historically, this river functioned as a distributary of the Teesta River and maintained a perennial flow. However, tectonic tilting in the Teesta basin—shifting drainage patterns toward the southeast—eventually severed this link, transforming the river into a monsoon-dependent system. Previously, it served as a vital channel for Teesta's discharge into the Ganges. A major hydrological transformation occurred in 1787 when the Teesta abandoned its original path and redirected its flow into the Brahmaputra, marking a significant shift in the riverine network. The river's hydrology is primarily governed by the seasonal rainfall of a subtropical monsoon

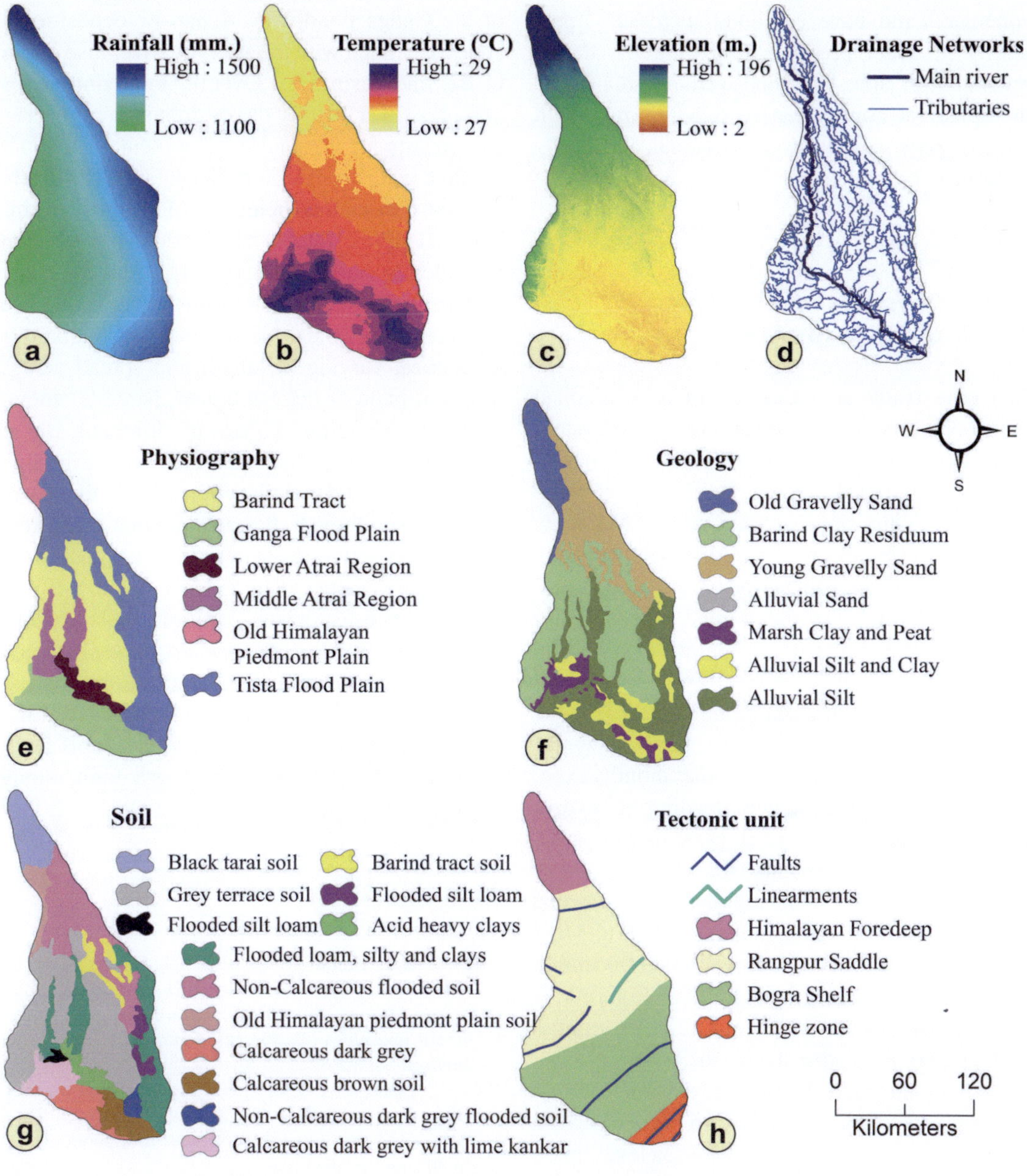

Fig. 1.6 Climate conditions and different physiographic aspects of Atreyee River basin **a** annual average rainfall, **b** temperature, **c** relief, **d** drainage network, **e** physiography, **f** geology, **g** soil, **h** tectonic

climate, marked by distinct wet and dry phases. The average annual precipitation in the basin is around 1400 mm, with nearly 80% falling during the monsoon months (June to September). Dry conditions prevail in the winter (December to February) and pre-monsoon (March to May) periods. In 2012, the construction of a dam represented another major turning point in the basin's hydrological history. This structural intervention has contributed to increased variability in downstream water availability. Numerous wetlands, primarily rain-fed or seasonally inundated by floodwaters, are situated along the lower reaches of the river's floodplain. These wetlands are highly ephemeral and have exhibited increased irregularity in water presence following the dam's regulation of flow. The total wetland area in the floodplain is approximately 628.27 km^2, with around 80% classified as seasonal, primarily sustained by monsoonal rains and episodic flooding.

1.5.1 Physical Characteristics of the Study Area

The case study area lies within a subtropical monsoon climatic zone, receiving approximately 1400 mm of rainfall annually. This precipitation is unevenly spread across four distinct climatic periods: (1) the pre-monsoon phase (March to May), (2) the monsoon phase (June to September), (3) the post-monsoon period (October to mid-December), and (4) the winter season (mid-December to February). A significant proportion of the total rainfall—around 80%—is concentrated during the monsoon months, primarily due to the influence of south-westerly monsoon winds. Conversely, the winter and summer months experience limited precipitation, leading to hydrological stress on both river systems and adjacent wetland environments (Rashid et al. 2013). Thermal conditions in the region exhibit seasonal extremes: summer temperatures (March to May) often surpass 40 °C, while winter lows (December to February) can drop below 10 °C. Figure 1.6a and b illustrates the annual variations in rainfall and temperature across the basin.

Topographically, the highest elevation within the river basin is located at Siliguri in the upstream zone, rising to 196 m above mean sea level (msl), while the lowest point—near the river's confluence—rests at just 2 m above msl (Fig. 1.6c). Geographically, the basin is positioned between the Ganga and Brahmaputra river systems and extends over parts of the Terai and Dooars in the north. This has given rise to varied physiographic characteristics across the basin (Fig. 1.6e). The northern section lies over the Old Himalayan Piedmont Plain, whereas the central and southern parts include the Teesta floodplain, Barind Tract with its ancient alluvial deposits, and portions of the Ganga floodplain. A narrow depositional zone associated with the Atreyee River also exists in the mid-basin area. Overall, low-lying zones comprise over half of the basin's area and serve as hotspots for wetland development. A distinct wetland belt is visible in the basin's lower stretch, with the Chalan Beel being one of the most prominent examples. Rashid et al. (2014) noted that tectonically induced sagging has played a key role in the development of these wetlands. The Atreyee basin is also networked with numerous tributaries, varying in length and spread across different parts of the catchment. Notable among them are the Talma, Fuleswari, Jorapani, Baral, Karatoya, Shib, Choto Jamuna, and Ichhmati rivers (Fig. 1.6d). Soil types within the basin are diverse, though floodplain soils deposited by riverine action dominate (Fig. 1.6g). From a geotectonic perspective, the basin lies in a region of active neo-tectonism. Multiple fault systems and linear tectonic features are present in and around the basin, indicating seismotectonic instability. Identified structures include the Katiha-Nilphamari fault, Gaibandha fault, Dhubri fault, Kardha fault, and Debgram-Bogra fault, along with several associated lineaments (Fig. 1.6h).

References

Almeida RM, Hamilton SK, Rosi EJ, Barros N, Doria CR, Flecker AS et al (2020) Hydropeaking operations of two run-of-river mega-dams alter downstream hydrology of the largest Amazon tributary. Front Environ Sci 8:120

Baba A, Tsatsanifos C, El Gohary F, Palerm J, Khan S, Mahmoudian SA et al (2018) Developments in water dams and water harvesting systems throughout history in different civilizations

Barbarossa V, Schmitt RJ, Huijbregts MA, Zarfl C, King H, Schipper AM (2020) Impacts of current and future large dams on the geographic range connectivity of freshwater fish worldwide. Proc Natl Acad Sci 117(7):3648–3655

Bartholomew MK, Anderson CJ, Berkowitz J (2019) Soil conditions following hydrologic restoration in cypress dome wetlands. Wetlands 39:185–196

Brunner P, Simmons CT, Cook PG (2009) Spatial and temporal aspects of the transition from connection to disconnection between rivers, lakes and groundwater. J Hydrol 376(1–2):159–169

Burkhard B, Kroll F, Nedkov S, Müller F (2012) Mapping ecosystem service supply, demand and budgets. Ecol Ind 21:17–29

Burkhard B, Kandziora M, Hou Y, Müller F (2014) Ecosystem service potentials, flows and demands-concepts for spatial localisation, indication and quantification. Landscape Online, 34

Chen W, Cao C, Liu D, Tian R, Wu C, Wang Y et al (2019) An evaluating system for wetland ecological health: case study on nineteen major wetlands in Beijing-Tianjin-Hebei region, China. Sci Total Environ 666:1080–1088

Church M (1995) Geomorphic response to river flow regulation: case studies and time scales. Regul Rivers Res Manage 11(1):3–22

Clarke B, Thet AK, Sandhu H, Dittmann S (2021) Integrating cultural ecosystem services valuation into coastal wetlands restoration: a case study from South Australia. Environ Sci Policy 116:220–229

Costanza R, De Groot R, Sutton P, Van der Ploeg S, Anderson SJ, Kubiszewski I et al (2014) Changes in the global value of ecosystem services. Glob Environ Chang 26:152–158

Davidson NC, Finlayson CM (2018) Extent, regional distribution and changes in area of different classes of wetland. Mar Freshw Res 69(10):1525–1533

Davidson NC, Van Dam AA, Finlayson CM, McInnes RJ (2019) Worth of wetlands: revised global monetary values of coastal and inland wetland ecosystem services. Mar Freshw Res 70(8):1189–1194

Davies BR, Beilfuss RD, Thoms MC (2000) Cahora Bassa retrospective, 1974–1997: effects of flow regulation on the Lower Zambezi River. InternationaleVereinigung Für Theoretische und angewandteLimnologie: Verhandlungen 27(4):2149–2157

De Groot R, Brander L, Van Der Ploeg S, Costanza R, Bernard F, Braat L, et al (2012) Global estimates of the value of ecosystems and their services in monetary units. Ecosyst Serv 1(1):50–61

Dey P, Saha SK, Chakraborty S (2020) Confluence of channel dimensions and groove width dictates slippery hydrodynamics in grooved hydrophobic confinements. Microfluid Nanofluid 24:1–15

Dungan PJ (ed) (1990) Wetland conservation: a review of current issues and required action

Everard M (2019) A socio-ecological framework supporting catchment-scale water resource stewardship. Environ Sci Policy 91:50–59

Fahlbusch H (2009) Early dams. Proc Inst Civ Eng Eng Hist Herit 162(1):13–18

Fergus T (1997) Geomorphological response of a river regulated for hydropower: River Fortun, Norway. Regul Rivers Res Manag Int J Devoted River Res Manag 13(5):449–461

Fluet-Chouinard E, Stocker BD, Zhang Z, Malhotra A, Melton JR, Poulter B et al (2023) Extensive global wetland loss over the past three centuries. Nature 614(7947):281–286

Fullerton AH, Burnett KM, Steel EA, Flitcroft RL, Pess GR, Feist BE et al (2010) Hydrological connectivity for riverine fish: measurement challenges and research opportunities. Freshw Biol 55(11):2215–2237

Germanovski D, Ritter DF (1988) Tributary response to local base level lowering below a dam. Rivers. https://doi.org/10.1002/rrr.3450020103

Graf WL (2006) Downstream hydrologic and geomorphic effects of large dams on American rivers. Geomorphology 79(3–4):336–360

Guo E, Chen L, Sun R, Wang Z (2015) Effects of riparian vegetation patterns on the distribution and potential loss of soil nutrients: a case study of the Wenyu River in Beijing. Front Environ Sci Eng 9:279–287

Gurnell AM, Downward SR, Jones R (1994) Channel planform change on the River Dee meanders, 1876–1992. Regul Rivers Res Manage 9(4):187–204

Hunt JD, Byers E, Wada Y, Parkinson S, Gernaat DE, Langan S et al (2020) Global resource potential of seasonal pumped hydropower storage for energy and water storage. Nat Commun 11(1):947

ICOLD C (ed) (2021) Position paper dam safety and earthquakes. CRC Press

Jiang W, Pan H, Yang N, Xiao H (2023) Dam inundation duration as a dominant constraint on riparian vegetation recovery. Sci Total Environ 904:166427

Knight JM, Dale P, Dwyer P, Marx S (2017) A conceptual approach to integrate management of ecosystem service and disservice in coastal wetlands

Kopperoinen L, Itkonen P, Niemelä J (2014) Using expert knowledge in combining green infrastructure and ecosystem services in land use planning: an insight into a new place-based methodology. Landscape Ecol 29:1361–1375

Koschke L, Fürst C, Frank S, Makeschin F (2012) A multicriteria approach for an integrated land-cover-based assessment of ecosystem services provision to support landscape planning. Ecol Ind 21:54–66

Kulcsár Z, Augsburger L, Reymond P, Pereira VM, Hirsch S, Mallik AS et al (2012) Flow diversion treatment: intra-aneurismal blood flow velocity and WSS reduction are parameters to predict aneurysm thrombosis. Acta Neurochir 154:1827–1834

Lai X, Chen H, Hou Y, Finlayson B, Li M, Chen J (2021) Lowering water level of Dongting lake of the Mid-Yangtze River in response to large-scale dam construction: a 60-year analysis. Geomorphology 391:107894

Larsen A, Larsen JR, Lane SN (2021) Dam builders and their works: beaver influences on the structure and function of river corridor hydrology, geomorphology, biogeochemistry and ecosystems. Earth Sci Rev 218:103623

Let M, Pal S (2023) Socio-ecological well-being perspectives of wetland loss scenario: a review. J Environ Manage 326:116692

Let M, Pal S, Let M, Ghosh R, Debanshi S (2024) Anthropogenic impact on ecosystem service value of urban blue space in Old Malda Municipality of eastern India. Environ Monit Assess 196(10):976

Li T, Zhang W, Zhang Q, Lu Y, Wang G, Niu Z et al (2015) Impacts of climate and reclamation on temporal variations in CH_4 emissions from different wetlands in China: from 1950 to 2010. Biogeosciences 12(23):6853–6868

Li J, Lai X, Liu H, Yang D, Zhang G (2018) Emergy evaluation of three rice wetland farming systems in the Taihu lake catchment of China. Wetlands 38:1121–1132

Lin W, Xu D, Guo P, Wang D, Li L, Gao J (2019) Exploring variations of ecosystem service value in Hangzhou Bay Wetland, Eastern China. Ecosyst Serv 37:100944

Liu D, Cui Y, Jin W, Wang H, Tang H (2023) Channel aggradation triggered by dam failure amplifies the damage of outburst flood. Landslides 20(7):1343–1362

Lu XX, Siew RY (2006) Water discharge and sediment flux changes over the past decades in the Lower Mekong River: possible impacts of the Chinese dams. Hydrol Earth Syst Sci 10(2):181–195

Lyons J (1992) The length of stream to sample with a towed electrofishing unit when fish species richness is estimated. North Am J Fish Manag 12(1):198–203

Ma C, Peng M, Zhang L, Shi Z, Zhou J, Chen H et al (2024) Erosion, deposition and breach evolution of landslide dams composed of various dam material types based on flume tests. Eng Geol 337:107598

Maithya JK, Ming'ate FL, Letema SC (2022) The impact of wetland utilisation on provisioning ecosystem services in Nyando Wetland, Kenya. J Environ Assess Policy Manag 24(01):2250023

McInnes RJ, Everard M (2017) Rapid assessment of wetland ecosystem services (RAWES): an example from Colombo, Sri Lanka. Ecosyst Serv 25:89–105

Millennium Ecosystem Assessment (2005) Living beyond our means: natural assets and human wellbeing. Island Press, Washington, DC

Mishra RK (2023) Fresh water availability and its global challenge. Br J Multidiscip Adv Stud 4(3):1–78

Mitsch WJ, Gosselink JG (2000) The value of wetlands: importance of scale and landscape setting. Ecol Econ 35(1):25–33

Mitsch WJ, Gosselink JG (2007) Wetlands, 4th edn. Wiley, Hoboken

Mitsch WJ, Nahlik A, Wolski P, Bernal B, Zhang L, Ramberg L (2010) Tropical wetlands: seasonal hydrologic pulsing, carbon sequestration, and methane emissions. Wetlands Ecol Manage 18:573–586

Mondal D, Pal S (2017) Evolution of wetlands in lower reaches of Bagmari–Bansloi–Pagla rivers: a study using multidated images and maps. Curr Sci 112(11):2263–2272

Montoya-Tangarife C, De La Barrera F, Salazar A, Inostroza L (2017) Monitoring the effects of land cover change on the supply of ecosystem services in an urban region: a study of Santiago-Valparaíso, Chile. PLoS ONE 12(11):e0188117

Morris JT, Shaffer GP, Nyman JA (2013) Brinson review: perspectives on the influence of nutrients on the sustainability of coastal wetlands. Wetlands 33:975–988

Munia H, Guillaume JHA, Mirumachi N, Porkka M, Wada Y, Kummu M (2016) Water stress in global transboundary river basins: significance of upstream water use on downstream stress. Environ Res Lett 11(1):014002

Olson LH, Misenheimer JC, Nelson CM, Bradham KD, Richardson CJ (2017) Influences of coal ash leachates and emergent macrophytes on water quality in wetland microcosms. Water Air Soil Pollut 228:1–11

Opperman JJ, Galloway GE, Fargione J, Mount JF, Richter BD, Secchi S (2009) Sustainable floodplains through large-scale reconnection to rivers. Science 326(5959):1487–1488

Pal R, Biswas SS, Pramanik MK, Mondal B (2016) Bank vulnerability and avulsion modeling of the Bhagirathi-Hugli river between Ajay and Jalangi confluences in lower Ganga Plain, India. Model Earth Syst Environ 2:1–10

Prasad SN, Ramachandra TV, Ahalya N, Sengupta T, Kumar A, Tiwari AK et al (2002) Conservation of wetlands of India-a review. Trop Ecol 43(1):173–186

Pritchard J, Lewis T, Beeching L, Denham T (2019) An assessment of microCT technology for the investigation of charred archaeological parenchyma from house sites at Kuk Swamp, Papua New Guinea. Archaeol Anthropol Sci 11:1927–1938

Quyen NTK, Berg H, Gallardo W, Da CT (2017) Stakeholders' perceptions of ecosystem services and Pangasius catfish farming development along the Hau River in the Mekong Delta, Vietnam. Ecosyst Serv 25:2–14

Rahimi L, Malekmohammadi B, Yavari AR (2020) Assessing and modeling the impacts of wetland land cover changes on water provision and habitat quality ecosystem services. Nat Resour Res 29(6):3701–3718

Ramsar Convention (1971) International conference on the conservation of wetlands and waterfowl, Ramsar, Iran, 30 January–3 February 1971. Final act and summary. Ramsar Convention, Gland, Switzerland

Rashid B, Islam SU, Islam B (2014) Drainage characteristics and evolution of the Barind tract, Bangladesh. Am J Earth Sci 1(4):86–98

Rashid MB, Islam MB, Sultan-Ul-Islam M (2013) Causes of acute water scarcity in the Barind Tract, Bangladesh. Int J Econ Environ Geol 4(1):5–14

Sarda R, Pal S (2022) Evaluating damming effect on eco-hydrological alteration in river and wetland using indicators of hydrological alteration. Geocarto Int 37(27):16132–16156

Space Applications Centre (SAC) (2011) National wetland atlas. SAC, Indian Space Research Organisation, Ahmedabad

Suchara I (2019) The impact of floods on the structure and functional processes of floodplain ecosystems. J Soil Plant Biol 1:44–60

Tan G, Chen P, Deng J, Xu Q, Tang R, Feng Z, Yi R (2019) Review and improvement of conventional models for reservoir sediment trapping efficiency. Heliyon 5(9):e02458

Thoms MC, Walker KF (1993) Channel changes associated with two adjacent weirs on a regulated lowland alluvial river. Regul Rivers Res Manage 8(3):271–284

US EPA (2009) Clean Water Act Section 404. http://water.epa.gov/lawsregs/guidance/wetlands/sec404.cfm

Vidal-Abarca Gutiérrez MR, Nicolás-Ruiz N, Sánchez-Montoya MDM, Suárez Alonso ML (2023) Ecosystem services provided by dry river socio-ecological systems and their drivers of change. Hydrobiologia 850(12):2585–2607

Vihervaara P, Rönkä M, Walls M (2010) Trends in ecosystem service research: early steps and current drivers. Ambio 39:314–324

Wondie A (2018) Ecological conditions and ecosystem services of wetlands in the Lake Tana Area, Ethiopia. Ecohydrol Hydrobiol 18(2):231–244

Wu W, Zhi C, Gao Y, Chen C, Chen Z, Su H et al (2022) Increasing fragmentation and squeezing of coastal wetlands: status, drivers, and sustainable protection from the perspective of remote sensing. Sci Total Environ 811:152339

Zedler JB, Kercher S (2005) Wetland resources: status, trends, ecosystem services, and restorability. Annu Rev Environ Resour 30(1):39–74

Damming and Downstream Channel Flow

2

Abstract

This chapter aimed to explore downstream change of diurnal, monthly, and yearly minimum, mean, and extreme channel flow pursuing damming using traditional to advance approaches. The study further tried to predict flow in the near future. Trend, variability, instability, periodicity, range of variability (RVA), failure flow estimation, machine learning (ML), etc., were applied to assess flow change and prediction. All types of flow assessment techniques reported declining minimum, mean, maximum, flow in dam downstream river. A marked reduction in peak flood discharge by a 66% in monsoonal flows was detected. In other seasons also the same trend was found. Increase of low-flow days and lowering of high-flow days in post-dam period were detected. Significant reduction of flood frequency in post-dam period is caused for ceasing water flow to the floodplain wetland and leading them in stress.

Keywords

Damming · Flow reduction · Flattening of peak flow · Dwindling flood · Reduction of flow variability · Rise of low-flow days

2.1 Dimensions of River Flow Modifications

Flow regime such as minimum flow, maximum flow, total flow, peakiness of flow, duration of peakiness, duration of low flow, base flow, inter-annual, and seasonal flow variability can control morphological and ecological characteristics of a river (Williams 1978), transport of nutrient, and sediment exchange (Poff et al. 1997). Ecological fabrics, its biodiversity, species richness, ecosystem structure (food chain or food web), etc., are majorly determined based on river flow regime (Bergkamp et al. 2000). Hence, the serviceability of river ecosystem also strongly depends on flow characteristics (Zeiringer et al. 2018). Since the flow regime and riparian floodplain processes are interwoven, change in flow regime could bring change in ecological ambience of the floodplain region concerned (Palmer and Ruhi 2019). Flow dynamics in river are also very natural, and system always tries to accommodate it with its self-regulatory homeostatic mechanism (Billman 2012). But massive change beyond tolerance limits often breaks the equilibrium condition of an ecological system which poses pressure on ecosystem functions and forms. Damming across

S. Pal et al., *Damming and Wetland Eco-hydrology*, GIScience and Geo-environmental Modelling,
https://doi.org/10.1007/978-3-032-11729-8_2

river primarily affects the hydrological system of the concerned river.

Dams not only alter the downstream flow regime but also upstream flow-form functions of rivers (Leopold et al. 2020; Liro 2014). The hydrological alteration by the dam construction may result in overall flow reduction, change in flow variability, base flow, reduction of peakiness, flood frequency and magnitudes, increase of low-flow days, decrease of high-flow days, and many others which collectively poses a great harm to river ecology (Petts 1984; Ward and Stanford 1979, 1995; Kondolf and Batalla 2005). This incident may further create longitudinal disconnection between upper and lower reaches of a river. This is merely not the hydrological disconnection, but also a disconnection of nutrient flow, sediment flow, ecological continuum, and so on (Chaudhari and Pokhrel 2022). Due to having changes in flow condition, in-channel geomorphological setup also changes, leading to changes in habitat condition.

In this potential scenario of hydrological changes in downstream river reach, hydrological poverty and ecological stress are very prominent. Further change may redirect the hydrological and ecological scenario. It is essential for comprehensive analysis of current course of changes and possible future changes. These are essential for river restoration and conservation planning purposes. This chapter tried to highlight the relevant methods that could be adopted for quantifying such changes. At the same time, the kind of changes that occurred or likely to be occurred is analyzed in reference to the Atreyee River.

2.2 Materials and Methods

Three hours of interval discharge and water level data from 1993 to 2024 collected from Joda bridge gauge station (lat. 25°13′30.03″N and long. 88°46′36.03″E) at Balurghat (under Irrigation and Waterways Dept., Govt. of West Bengal) over Atreyee River was used for analyzing its changing hydrological regime from very micro- to macrotemporal resolutions.

2.3 Methods of Measuring Flow Alteration

2.3.1 Change Point Assessment

Identifying abrupt shifts or breakpoints in climatic and hydrological time series can be achieved using several statistical techniques. Among the widely recognized methods are the Pettitt test (Pettitt 1979) and the Standard Normal Homogeneity Test (SNHT) (Alexandersson and Moberg 1997), both commonly used to assess changes in river discharge patterns. While these specific tests were not applied in this study—due to the clear disruption caused by the construction of a dam on the Atreyee River in 2010—they are briefly explained here for academic reference.

Pettitt Test: This method is a nonparametric, rank-based approach designed to detect a significant shift in the central tendency (mean) of a time series. Notably, it does not require a predefined hypothesis regarding the location of the change point. It has been extensively utilized in the analysis of hydrological and climatic datasets (Zhang and Lu 2009; Gao et al. 2011). Let **t** represent the full length of the series, with a potential shift at time **m**. The test statistic is computed using Eq. 2.1. The statistic is similar to the Mann–Whitney statistic, which characterized by two samples, such as y_1, $y_2, \ldots, y_m$ and $y_m + 1, y_2, \ldots, y_n$:

$$U_{t,m} = \sum_{i=1}^{m} \sum_{j=t+1}^{t} \operatorname{sgn}(y_i - y_j), \tag{2.1}$$

where sgn in Eq. 2.1 is defined by Eq. 2.2:

$$\operatorname{sgn}(y_i - y_j) = \begin{pmatrix} 1 & \text{if}(y_i - y_j)\,\rangle 1 \\ 0 & \text{if}(y_i - y_j) = 0 \\ -1 & \text{if}(y_i - y_j)\,\langle 1 \end{pmatrix}. \tag{2.2}$$

The statistic $U_{t,m}$ is calculated across all random observations within the range of 1 to n. The most probable point of change is identified where the absolute value $|U_{t,m}|$ reaches its maximum, as described in Eq. 2.3.

$$K_T = \text{Max}_{1 \leq t \langle m} |U_{t.m}|. \tag{2.3}$$

The probability of shifting year is estimated when $|U_{t,m}|$ is maximum following Eq. 2.4:

$$P = 1 - \exp\left(\frac{-6k_T^2}{y^2 + y^3}\right). \tag{2.4}$$

A *p*-value below the significance threshold α\alphaα suggests rejection of the null hypothesis.

Standard Normal Homogeneity Test (SNHT): Also known as the Alexanderson test, the SNHT is suitable for detecting step changes or inhomogeneities in time series, especially for hydrometeorological records. The test statistic is calculated following Eq. 2.5:

$$T_s = \max T_m,\ 1 \leq m \langle n, \tag{2.5}$$

where T_s attains maximum value in the data series that is called change point and T_m could be derived using Eq. 2.6:

$$T_m = \overline{m}z_1 + (n - m))\overline{z}_1,\ m = 1, 2, \ldots, n \tag{2.6}$$

where

$$\overline{z}_1 = \frac{1}{m} \sum_{i=1}^{n} \frac{(M_i - \overline{M})}{s} \tag{2.7}$$

where $\overline{M}$ represents mean and s represents the standard deviation of the sample data.

Trend Analysis

Mann-Kendall test: The Mann-Kendall (MK) test (Mann 1945; Kendall 1975) is a rank-based, nonparametric method for trend detection in hydrologic and climatic time series (Yue and Wang 2002; Yue et al. 2002). It is particularly robust in handling non-normal datasets and is less influenced by sudden discontinuities. The analysis here was conducted using the R software environment.

Prior to applying the MK test, it is crucial to assess and remove any serial correlation in the dataset, as positive autocorrelation may falsely indicate significant trends (Jenkins 1968; von Storch and Navarra 1999). To address this, Trend-Free Pre-Whitening (TFPW) as described by Yue et al. (2002) was employed to eliminate serial correlation effects.

To estimate the slope or rate of change in a dataset, the Theil–Sen method (Sen 1968) was applied. This approach calculates the median of all possible pairwise slopes using Eq. 2.8:

$$T_i = \text{Median}\frac{X_m - X_n}{m - n}. \tag{2.8}$$

For all $m > n$ where X_m and X_n are values of the variable at times m and n, respectively. In addition to Theil–Sen, linear regression may also be used to estimate trend significance and slope (Tangang et al. 2007; Schaefer and Domroes 2009).

The percentage change over a period: To assess the percent change in streamflow (annually or seasonally), the trend slope is estimated based on the natural logarithm of flow data. Sen's slope estimator is then applied to the logged data (Helsel and Hirsch 1993; Lehmann and Totsche 2020) and calculated using Eq. 2.9.

$$\text{Percentage change (\%) over period} = \left(\frac{\text{Median slope} \times \text{Length of the period}}{\text{Mean}}\right) \tag{2.9}$$

Spearman's Rank correlation Coefficient (rho): To quantify the degree of monotonic relationship between two variables, Spearman's rank correlation is used. It can be calculated following Eq. 2.10.

$$r_s = \frac{1 - 6\sum_{i=1}^{n} t_i^2}{n^3 - n}. \tag{2.10}$$

Here, *ti* is the difference in ranks between paired observations *Xi* and *Yi*, and **n** is the number of data pairs.

For small sample sizes, the statistical significance of ρ can be obtained from standard critical value tables. When n>20, a *t*-distribution approximation can be used for hypothesis testing.

2.3.2 Methods for Quantifying Hydrological Alteration

Figure 2.1 summarizes the methods that could be adopted for quantifying hydrological alteration. Some frequently used conventional approaches are flow analysis, flood analysis, base flow analysis, regional hydrologic analysis, etc. Along with this, some advanced methods like range of variability (RVA), degree of hydrological alteration, machine learning approaches, etc., could also be used fruitfully.

2.3.2.1 Degree of Impoundment

To evaluate the extent of flow regulation by a dam, the impoundment ratio (IR) is utilized. This metric is defined as the ratio between a dam's storage volume and the total water yield from its contributing watershed, resulting in a unitless value. In this assessment, storage was calculated using dead storage to avoid overestimating the hydrological influence—using total storage volume may exaggerate dam-related impacts (Batalla et al. 2004). A low IR indicates minimal control over natural flow regimes, while a higher IR reflects stronger regulation. Dams with relatively small reservoirs typically have limited hydrological influence. However, in some cases, continuous water diversion through canals may reduce the IR numerically, despite the actual field impact being more substantial.

2.3.2.2 Flood Frequency and Magnitude

To investigate the influence of damming on flood dynamics, flood frequency analysis can be performed using statistical methods such as the Gumbel Type I distribution or Pearson Type III distribution. In addition to these models, a simpler approach involves comparing historical water levels against danger level (DL) and extreme danger level (EDL) thresholds. If water levels surpass the EDL, flooding is assumed to have occurred. Graphing annual water levels in relation to the EDL allows for a count of years in which floods were triggered. Furthermore, the difference between actual water levels and the EDL can serve as a proxy for flood magnitude. For this study, the EDL-based method was adopted to analyze flood frequency before and after dam construction.

2.3.2.3 Estimation of Base Flow

Base flow was estimated for both pre- and post-dam scenarios using the straight-line separation

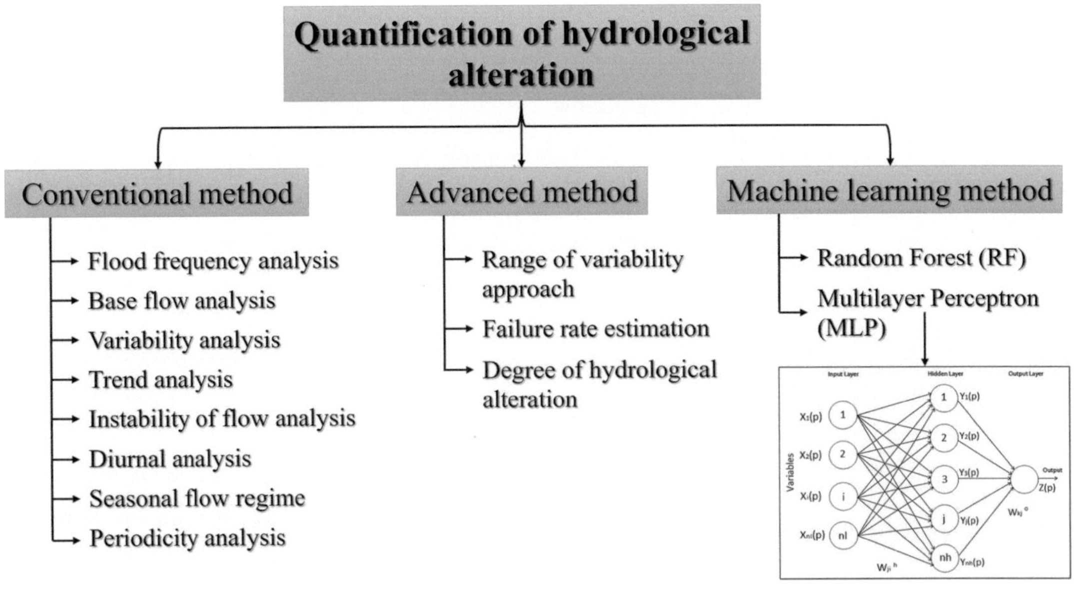

Fig. 2.1 Techniques used for the quantification of hydrological alteration

method as described by Linsley et al. (1975). This technique involves drawing a line on a discharge hydrograph from the point where runoff begins to decline to the end of high flow, where the contribution transitions predominantly to groundwater. The segment of the hydrograph beneath this line represents the base flow, while the portion above reflects surface runoff. This method helps in delineating changes in groundwater discharge due to dam-induced modifications in streamflow patterns.

2.3.2.4 Measuring Seasonal Flow Regime

Entire time spectrum was classified into pre-monsoon (March to May), monsoon (June to September), and post-monsoon (October to December) flow regimes. Discharge or water level (WL) data of pre- and post-dam times was illustrated on line graphs showing time series variation of seasonal flow following Batalla et al. (2004) and Graf (2006). Monthly average maximum, minimum flow could be plotted. Average situation of the same for pre- and post-dam regimes could also be plotted in the form of hydrograph at a same scale in order to see the discharge gap. Heat map is another approach that could be adopted for presenting year-wise maximum, minimum, and average discharges continuum over time. It could also help to introspect the continuity break (if any) within the entire time spectrum in reference to dam. Rate of change and its continuum is also an approach that could also be fruitfully applied to present the degree of discharge change scenario. Both these approaches were adopted for the present case.

2.3.2.5 Measuring Flow Variability, Trend, and Instability Analysis

To understand how river discharge trends have shifted before and after dam construction, linear regression analysis using the least squares method was applied, with the coefficient of determination (R^2) used to quantify the goodness of fit. Additionally, the coefficient of variation (CV) for flow parameters such as mean, maximum, and minimum discharge was calculated to assess variability. Changes in monthly flow regimes between pre- and post-dam periods were identified using the ratio of average, maximum, and minimum monthly flows. A ratio equal to 1 implies no change; values greater than 1 suggest a reduction in peak flows, while values less than 1 point to an increase in base flow or surface runoff under regulated conditions.

Flow instability—indicative of fluctuations in the time series such as streamflow or water level—was examined using the Instability Index (IX) developed by Cuddy and Della Valle (1978). This metric combines both temporal variation and trend, making it suitable for evaluating dynamic discharge behaviors. The IX is defined by Eq. 2.11:

$$\mathrm{IX} = \mathrm{CV} \times \sqrt{1 - R^2} \tag{2.11}$$

where IX is the instability index, CV represents the coefficient of variation, and R^2 is the coefficient of determination. Lower values of IX indicate more stable conditions, whereas higher values denote increased instability in discharge patterns.

2.3.2.6 Diurnal Flow Frequency Analysis

To explore variations in daily discharge patterns, flow duration curves (FDCs) were constructed, which plot mean daily discharge against the probability of exceedance. Additionally, a frequency distribution of diurnal flows across selected discharge classes was generated separately for the pre- and post-dam periods. This graphical method, though straightforward, provides meaningful insights into changes in short-term flow behavior due to regulation and was adopted in this analysis.

2.3.2.7 Methods for Regional Hydrologic Analysis

Understanding the spatial impacts of damming at a regional scale involved calculating metrics such as the maximum-to-mean and maximum-to-minimum annual discharge ratios. These ratios reflect the extent of flow variability: higher values signal greater fluctuations, typically associated with large floods and geomorphologically complex rivers, while lower values suggest steady flow and more uniform channel forms. The maximum/mean flow ratio for each year is particularly insightful, indicating how flood peaks compare to average conditions. To measure relative changes in flow due to regulation, the method of standardized percentage change (Pc) as proposed by Benn and Erskine (1994) was employed:

$$\text{Pc} = 100 \times \left(\frac{X_2 - X_1}{Y_2 - Y_1} \right) - 1 \qquad (2.12)$$

where Pc is the standardized percentage change of mean daily discharge of a given duration. Here, X_1 and X_2 are the mean daily discharges at two sites under natural (unregulated) conditions, while Y_1 and Y_2 represent the flows under regulated (post-dam) conditions. This method is particularly useful when multiple monitoring stations are available along the same river system.

For a more comprehensive understanding of flow regime shifts, several hydrological indicators recommended by Graf (2006) were examined. These include metrics like 1-day and instantaneous maximum/minimum flows, their timing, flow pulse counts and durations, and ramping rates, offering detailed insights into hydrological responses to regulation.

2.3.2.8 Methods for Analyzing Threshold Discharge and Degree of Impact

Flow alterations were further assessed by comparing pre- and post-dam flows to established thresholds. One robust method for identifying such thresholds is the range of variability approach (RVA), originally developed by Richter et al. (1997) and later refined by Smakhtin et al. (2006) and Monk et al. (2007). The RVA is designed to evaluate deviations in flow magnitude, frequency, duration, and timing. For this study, standard deviation-based RVA was applied to monthly maximum and minimum discharges using Eq. 2.13.

$$(\text{mean} - \text{SD}) \leq \text{parameter} \leq (\text{mean} + \text{SD}). \qquad (2.13)$$

To quantify the overall degree of hydrologic alteration, the Euclidean distance method developed by Dyer et al. (2014) was used. It summarizes deviations across multiple flow indicators:

$$\text{IHA} = \frac{1}{100}\sqrt{\left[\frac{1}{n}\sum_{n}(I_i)^2\right]}\forall_j. \qquad (2.14)$$

where IHA_j denotes the overall hydrologic alteration index for region j, I_i is the absolute percent deviation of indicator iii, and nnn is the number of indicators. A value near 0 implies minimal alteration, while values close to or above 1 reflect significant flow modification.

2.3.2.9 Flow Simulation and Forecasting

To simulate and predict streamflow under regulated conditions, machine learning (ML) techniques were implemented. Specifically, multilayer perceptron (MLP) and random forest (RF) algorithms were selected due to their demonstrated accuracy in hydrological modeling.

The MLP, a type of artificial neural network (ANN), includes an input layer, one or more hidden layers, and an output layer. Neurons within the network are interconnected with weighted links that adjust during training to optimize predictions (Pourghasemi et al. 2020). The training process is governed by the back-propagation algorithm, which iteratively reduces prediction error (Zare et al. 2013; Zhang et al. 2020). Activation functions like ReLU and sigmoid introduce nonlinear capabilities, making MLP suitable for modeling complex hydro-climatic patterns (Khatun et al. 2021).

On the other hand, the random forest (RF) algorithm, introduced by Breiman (2001), constructs an ensemble of decision trees using bootstrap samples and randomly selected predictor variables at each split. The method improves over traditional bagging techniques (Polikar 2012) and does not assume any distributional form of input

Table 2.1 Calculated parameters for different algorithms used in the study

Parameters	ANN-MLP	RF
Lags		
Number of hidden layer	10	–
Seed	2	1
Number of iteration	1000	100
Learning algorithm	Back-propagation	Random tree
Learning rate	0.3	–
Momentum	0.24	–
Depth of the tree	–	Unlimited
Size of bag	–	100
Batch size	100	100

data (Francke et al. 2008). Key model parameters include the number of trees (ntree), the number of variables considered at each node (mtry), and the minimum node size (nodesize). Final predictions are made by averaging the outputs of all trees in the forest.

To identify the most effective configuration for each predictive model, a trial-and-error approach was used to fine-tune input variables and model parameters. The finalized and optimized settings for both algorithms are summarized in Table 2.1.

2.3.2.10 Evaluation of the Models

To evaluate the predictive accuracy of the applied models, a range of standard statistical error metrics was used. These include Root Mean Square Error (RMSE), Mean Absolute Error (MAE), and Mean Absolute Percent Error (MAPE). These metrics are widely recognized for assessing the quality of predictions in hydrological modeling. RMSE measures the square root of the average of squared differences between observed and simulated values, and a lower RMSE value indicates a better-fitting model. MAE calculates the average of the absolute differences, offering a straightforward measure of prediction accuracy without emphasizing outliers. MAPE, on the other hand, expresses the error as a percentage, which helps in understanding the model's performance relative to the magnitude of observed values. In addition to these error metrics, the correlation coefficient was used to quantify the strength and direction of the linear relationship between the observed data and model outputs. A high correlation coefficient value, especially when statistically significant, suggests strong agreement between simulated and actual values and reflects the effectiveness of the model in capturing the dynamics of the system.

For this study, river discharge records from 1993 to 2024 were employed for simulation and forecasting. The reliability of the models was found to be strongly influenced by the proper calibration and optimization of their internal parameters. Careful tuning during the training process contributed significantly to the accuracy and robustness of the predictions (Fig. 2.2).

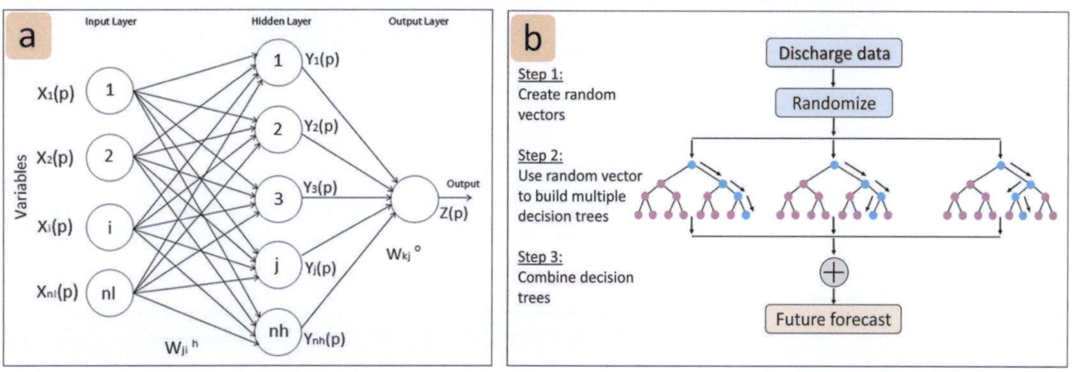

Fig. 2.2 Schematic structure of **a** ANN-MLP **b** random forest

Table 2.2 Number of year count under six discharge classes in monsoon months having a maximum discharge

Months	Discharge in cumec.					
	< 100	100–200	200–300	300–400	400–500	> 500
June	–	2000	–	–	–	1995
July	2012, 2013, 2014	2002, 2006, 2011	2010, 2021, 2022, 2023, 2024	1999	1998, 2003	1997, 2017
August	2015, 2016	2009	2008, 2019, 2020	–	1993, 1994	–
September		2004, 2007, 2018			1996	
October	–	–	1994, 2001	2005	–	–
Total	5	8	10	2	5	3

2.4 Result

2.4.1 Degree of Impoundments Analysis

Alive total water-holding capability of the dam is 7290000 m^3, and computed mean water delivered from the watershed is 3018093563 m^3. Computed IR index of this basin is very little (0.002415), which hypothetically signifies no such significant impact of dam on downstream river. But in ground reality, since the water is directly diverted through the canal network system and thereby controls downstream flow regime, here this value may not be very meaningful.

2.4.2 Reduction of Flood Peaks

The modal discharge-based monthly discharge distribution clarified that in post-dam period, and frequency of years with higher discharge was reduced significantly. Most of the post-dam years recorded modal discharge < 200 m^3/s during monsoon season. Contrarily, in pre-dam situation, eight years recorded discharge > 400 m^3/s (Table 2.2).

Peak water level plot in reference to extreme danger level (EDL) (23.76 m) showed that 46% years in pre-dam period recorded flood situation but in post-dam period, water levels in all the years (except 2017) couldn't reach even the primary danger level (PDL) (Fig. 2.3).

2.4.3 Annual and Seasonal Discharge Reduction Scenario

After construction of the dam, Atreyee River experienced a significant reduction of about 59% dwindling of downstream discharge. Seasonally scenario is to some extent different. In pre-monsoon, monsoon, and post-monsoon seasons, about 16%, 66%, and 55% of discharges were reduced, respectively.

Fig. 2.3 Yearly maximum discharge from 1993 to 2014 showing discharge level above PDL, DL, and EDL

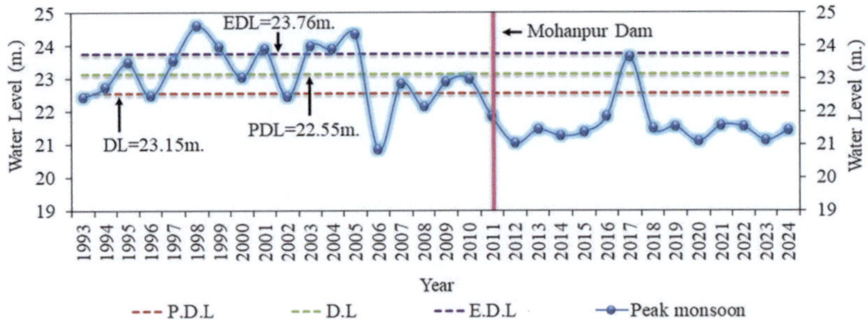

2.4.4 Base Flow Modification

Low flow often gets largely controlled by damming, since the demand for water is usually high during the lean season. Even a channel would be absolutely dry during lean season if base flow ceases its supply even with small water diversion (Batalla et al. 2004). In the present case, the volume of average base flow was eased about 18% (26 to 22 m^3/s) in post-dam period.

2.4.5 Seasonal Flow Regime Shift

Average monthly flow, maximum and minimum flow of pre- and post-dam periods are presented in Fig. 2.4. All these cases, it was observed that

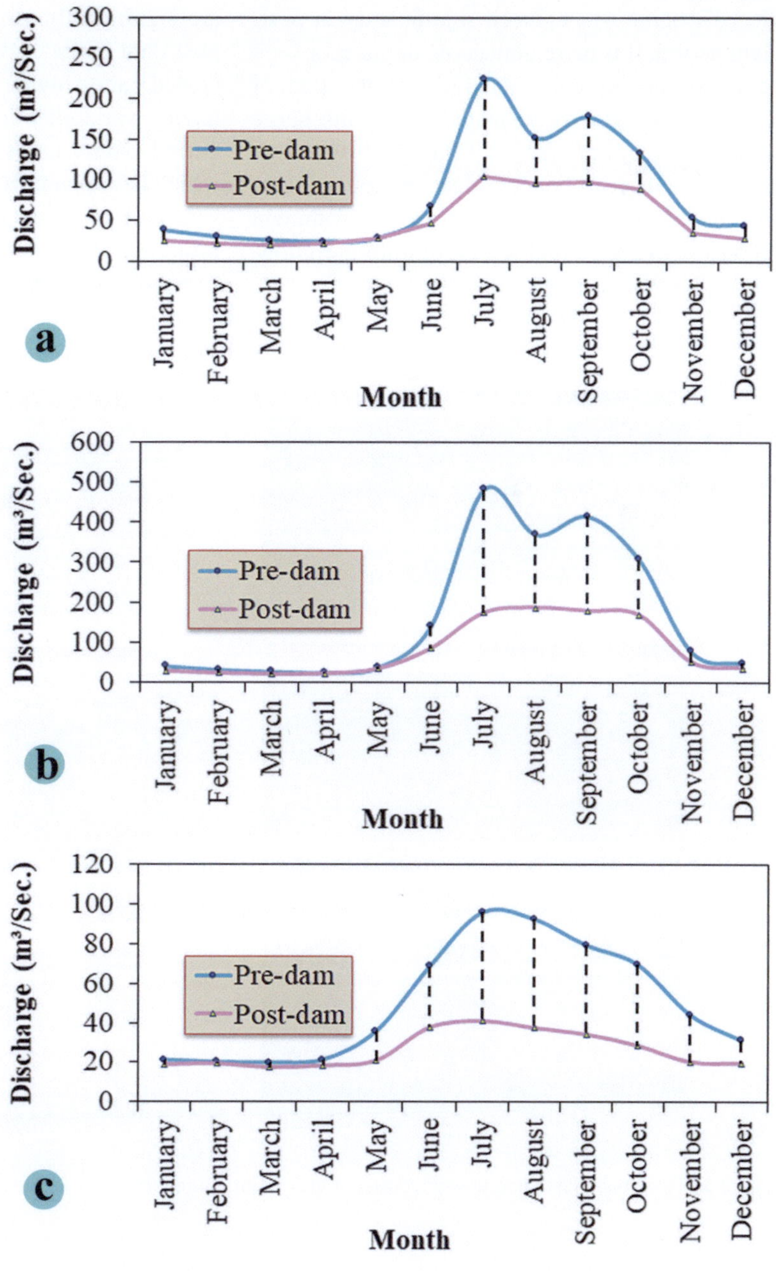

Fig. 2.4 **a** Average monthly, **b** average maximum, and **c** average minimum discharge before (1994–2012) and after (2013–2024) dam

peakiness of the hydrographs was eased in post-dam period. This rate is found rather high in case of maximum discharge. This signifies that role of damming toward flood control is very effective. Similar sorts of findings were also reported in the works of Guo et al. (2012). Not only was that inconsistency of discharge also declined considerably.

From the heat map of maximum, minimum, and average discharge and rate of change of discharge presented in Fig. 2.5, it is very evident that discharge was reduced significantly in post-dam period. It is more noticeable in pre-monsoon and monsoon months. Analysis of the rate of change also clearly depicted peak discharge change as more prominent. Change of discharge in different years of pre-dam period is also accounted, but adverse change is more prominent in post-dam period. In the case of minimum flow, the change rate is more contrasting.

2.4.6 Seasonal Flow Variability, Trend, and Instability

Flow variability of average, maximum, and minimum discharges computed based on CV is given in Table 2.3. Table clearly displays significant reduction of variability in all three cases in post-dam period. Box plot of discharge during monsoon and non-monsoon seasons separately showed the degree of dispersion of flow (Fig. 2.6). Both the volume of discharge and its dispersion are significantly greater in pre-dam period than post-dam period. Overall degree of dispersion is considerably higher in monsoon season due to having variability in monsoon rain. In pre-dam period, some cases of outliers were identified which has become almost nil in post-dam period.

Trend of seasonal average, maximum, and minimum flow analyses using linear least square regression were analyzed based on pre- and post-

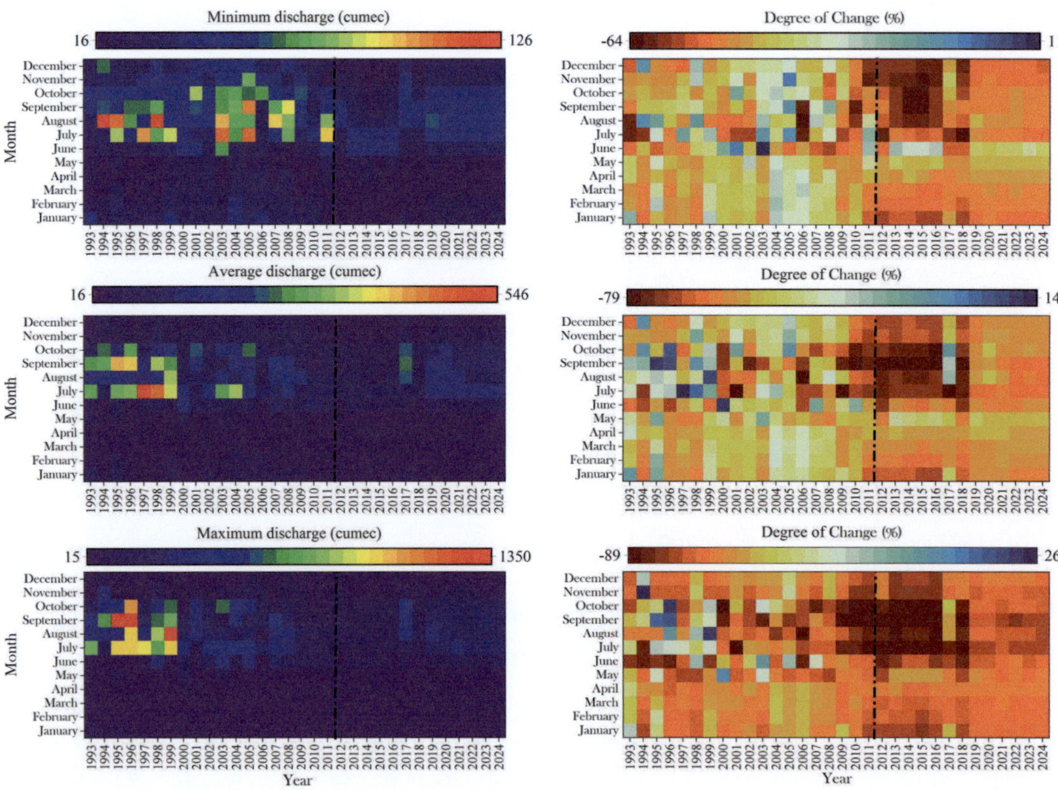

Fig. 2.5 Heat map illustrating flow alteration of maximum, minimum, and average discharge and their rate of changes

Table 2.3 CV of maximum, average, and minimum monthly discharge pre- and post-dam period

Month	CV of maximum discharge (%)		CV of average discharge (%)		CV of minimum discharge (%)	
	Pre-dam	Post-dam	Pre-dam	Post-dam	Pre-dam	Post-dam
January	38.89	2.75	27.43	5.102	29.03	2.96
February	27.60	4.93	22.97	0.49	20.80	2.65
March	21.70	5.36	52.63	2.97	19.46	2.95
April	24.23	12.22	19.00	3.28	17.56	7.08
May	64.48	37.90	26.73	42.17	16.71	9.77
June	179.80	35.77	134.47	29.05	45.02	33.87
July	64.14	3.613	65.00	15.18	45.28	26.11
August	82.58	18.38	54.68	16.81	41.29	19.10
September	99.22	9.44	67.64	12.81	62.06	35.38
October	84.49	11.04	56.56	8.87	51.61	21.16
November	73.54	45.71	51.23	8.87	42.71	9.123
December	36.77	39.41	32.90	13.67	29.01	0.77

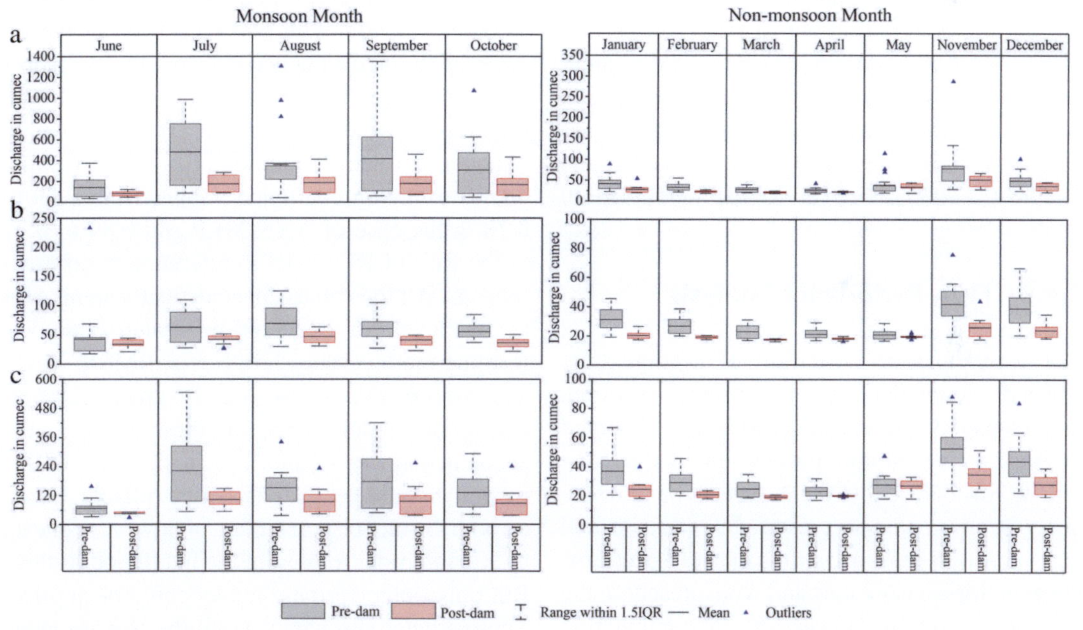

Fig. 2.6 Box plot showing comparative pattern of dispersion of maximum, minimum, and mean discharge data in pre- and post-dam periods in monsoon and non-monsoon months

dam time series datasets. In most of the cases, the overall trend is negative but always the degree of change is not statistically significant. Coefficient of determination (R^2) and concerned probability value computed established that. But trend analysis for entire time spectrum shows sharper change. Instability analysis in pre- and post-dam periods also revealed that instability of average, minimum, and maximum discharge in post-dam period was significantly declined (Table 2.4). Majority of the cases instability has become almost half.

Table 2.4 Season-specific instability (IX) shows time series change in average discharge in pre- and post-dam periods

Flow type	Season	Pre-monsoon	Monsoon	Post-monsoon
Average	Pre-dam	23.65	71.15	76.59
	Post-dam	9.63	32.53	23.41
Maximum	Pre-dam	49.85	98.81	142.92
	Post-dam	38.68	20.75	39.40
Minimum	Pre-dam	18.38	59.30	51.09
	Post-dam	9.26	25.13	21.35

2.4.7 Diurnal Flow Analysis

Figure 2.7 depicts the season-specific diurnal flow conditions in pre- and post-dam periods. In most of the cases, frequency of low daily flow was significantly enhanced. For example, in pre-monsoon season, about 58% days was accounted with a discharge between 20 and 30 m^3/s in pre-dam period but in post-dam period, 64.81% diurnal frequencies accounted discharge < 20 m^3/s. In monsoon season, frequency of days with a volume of discharge below 100 m^3/s was increased from 53.51 to 97.45%. These results witness for noticeable hydrological modification.

2.4.8 Flow Periodicity Analysis

The wavelet power spectrum of average time series discharge data in pre-monsoon, monsoon, post-monsoon, and winter seasons was presented in Fig. 2.8. In post-dam period, remarkable periodicity was detected in the 3–5 years band, especially in pre-monsoon season. In monsoon and post-monsoon seasons, strong power spectrums are at multiple year bands and were present in the pre-dam period. But prevalence and continuation are stronger in post-dam period. In winter season also, post-dam continuation of strong power spectrum between 3 and 5 years band is prominent. Although not presented here, such stronger periodicity was also detected in the case of minimum and maximum discharges.

2.4.9 Simulated and Predicted Flow Condition

Two machine learning models (ANN-MLP and RF) applied for predicting seasonal flow. Figure 2.9 depicts the season-wise observed and predicted flow up to 2024 in order to justify their applicability. Both the model outputs show parity of predicted flow with observed flow. However, in order to justify the better fit model, Pearson's correlation coefficient between observed and predicted flow was computed. The model with stronger correlation value signifies outperforming one. Correlation coefficient ranges from 0.69 to 0.78 in the case of ANN-MLP and 0.89 to 0.97 in the case of RF model in reference to different seasons. All the values are statistically significant at 0.01 level. So, RF could be treated as outperforming model (Table 2.5). Lower RMSE, MAE, and MAPE values also signify greater acceptability of model. Since all these values were accounted lower (Table 2.6) in the case of RF model, it could be treated as more reliable. Based on both the models, prediction of flow was done up to 2024 in order to present the likelihood of flow. But only outperforming model could be applied. The predicted discharges in all the seasons show the likelihood of discharge lowering (Fig. 2.10).

2.5 Discussion

The findings of this study reveal significant hydrological and ecological modifications induced by damming on the Atreyee River, particularly in terms of flow attenuation, seasonal variability, and base flow regulation. A marked reduction in peak flood discharge, as evidenced by a 66%

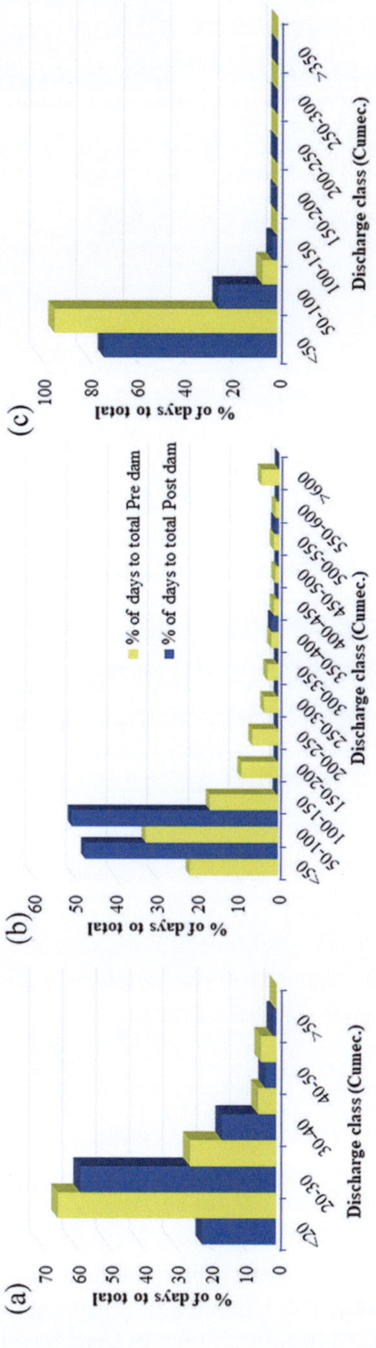

Fig. 2.7 Daily frequency of flow during **a** pre-monsoon, **b** monsoon, and **c** post-monsoon months

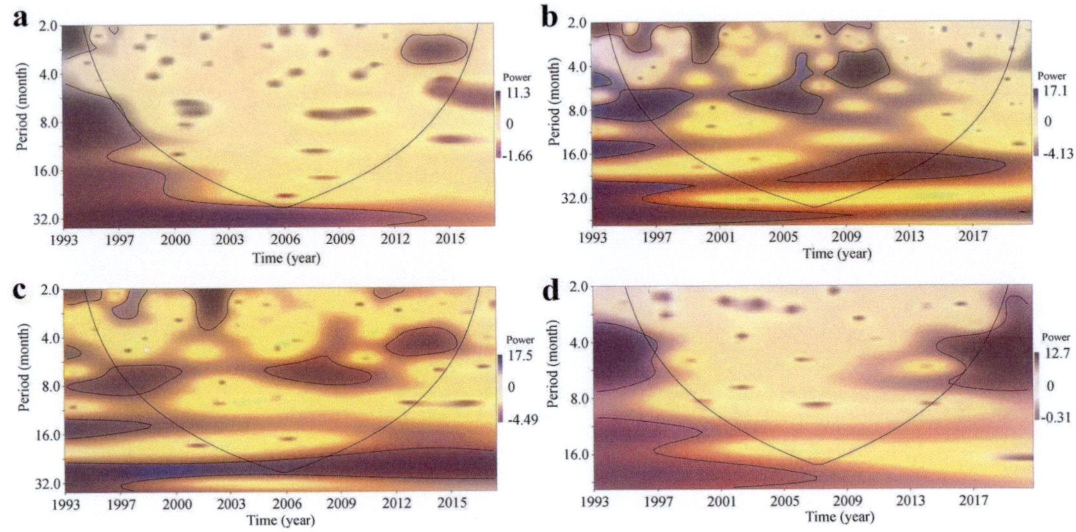

Fig. 2.8 **a–d** Wavelet power spectrum shows time series change in average discharge data from 1993 to 2024 for **a** pre-monsoon, **b** monsoon, **c** post-monsoon, and **d** winter seasons

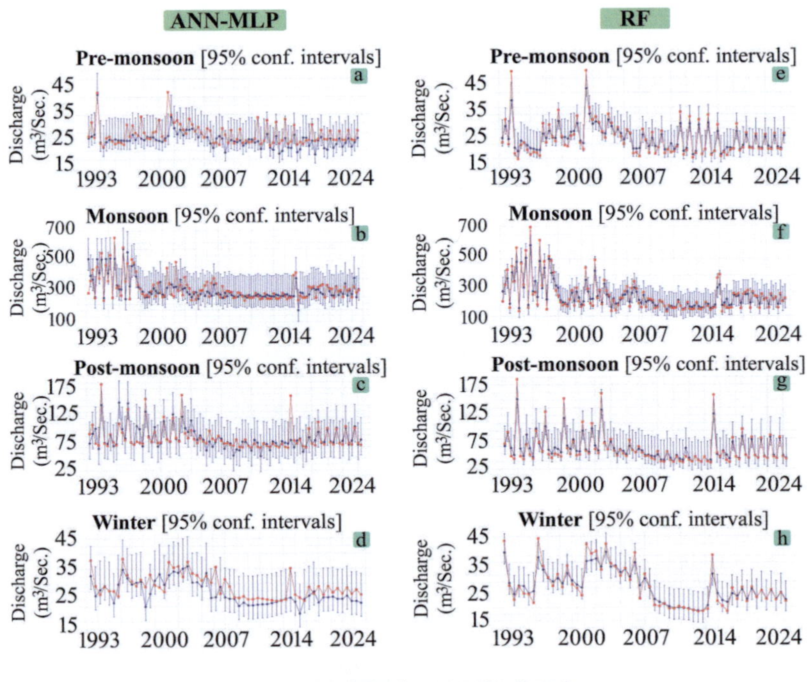

Fig. 2.9 River flow simulation for 1993–2024 using ANN-MLP and RF for different seasons: **a** and **e** pre-monsoon, **b** and **f** monsoon, **c** and **g** post-monsoon, and **d** and **h** winter

decrease in monsoonal flows, suggests a significant alteration in the natural flood regime. This has far-reaching implications for sediment transport, nutrient cycling, and riparian ecosystem stability (Poff et al. 1997; Ward and Stanford 1995). For instance, Guo et al. (2012) reported a sharp decline in flood magnitude and frequency in the Yangtze River following the construction of the Three Gorges Dam, leading to decreased sediment transport and disruptions in wetland hydrodynamics. Similarly, Batalla et al. (2004) documented significant flow regime alterations in the Ebro River,

Table 2.5 Pearson's correlation coefficient (*r*) between observed and simulated river flow and level of significance (sig.)

Models	Pre-monsoon		Monsoon		Post-monsoon		Winter	
	r	Sig.	*r*	Sig.	*r*	Sig.	*r*	Sig.
ANN-MLP	0.77	0.001	0.78	0.001	0.74	0.001	0.69	0.01
RF	0.97	0.01	0.91	0.01	0.89	0.05	0.92	0.05

Table 2.6 Computed RMSE, MAPE, and MAE errors between observed and simulated flow data for the used models

Models	Pre monsoon			Monsoon			Post monsoon			Winter		
	RMSE	MAPE	MAE	RMSE	MAPE	MAE	RMSE	MAPE	MAE	RMSE	MAPE	MAE
ANN	2.34	4.33	1.93	4.74	7.37	3.05	4.93	6.42	3.09	3.87	5.95	2.42
RF	2.01	3.71	1.44	3.98	6.61	2.52	3.49	5.53	2.78	2.45	5.84	1.72

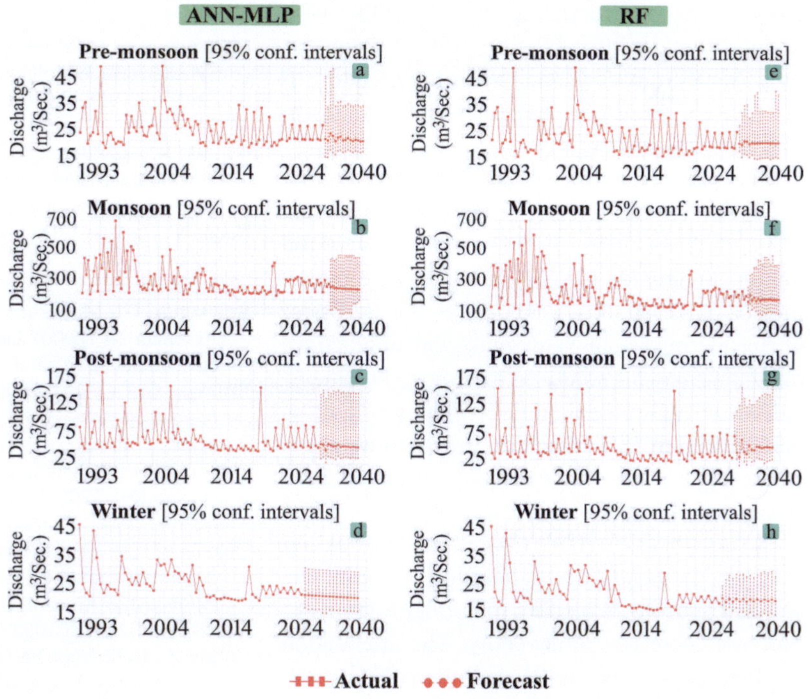

Fig. 2.10 River flow forecasting for 2024–2040 using ANN-MLP and RF for different seasons **a** and **e** pre-monsoon, **b** and **f** monsoon, **c** and **g** post-monsoon, and **d** and **h** winter

Spain, where reservoir operation not only reduced peak flows but also led to prolonged low-flow conditions, a trend evident in the Atreyee River post-damming. The observed decline in discharge variability, especially in monsoon months, aligns with previous studies indicating that flow regulation disrupts natural flow periodicity and impacts floodplain connectivity (Palmer and Ruhi 2019). The reduction in flow variability in the Atreyee River is also good tune with findings from Kondolf and Batalla (2005), who emphasized that dam-induced hydrological stability can disrupt sediment fluxes, altering downstream channel morphology and riparian habitats. The decrease in base flow by 18% in this study is comparable to the findings of Ward and Stanford (1995), who highlighted that flow regulation leads to a loss of seasonal hydrological pulses essential for sustaining wetland-dependent species. Moreover, studies by Richter et al. (1997) and Zeiringer et al. (2018) suggest that reduced hydrological connectivity, as observed in the Atreyee River, results in habitat fragmentation and biodiversity loss, particularly affecting species dependent on seasonal floodplain inundation. The ecological implications of these changes are also supported by the works of Palmer and Ruhi (2019), who emphasized that disruptions in natural flow regimes can lead to cascading effects on ecosystem structure and function. The decline in sediment deposition, an essential process for wetland fertility, has been similarly observed in studies on the Amazon River Basin, where Chaudhari and Pokhrel (2022) noted that hydropower dams disrupt the sediment continuum, thereby affecting nutrient cycling and primary productivity. This study further confirms observations by Bergkamp et al. (2000), who found that altered flood regimes due to damming significantly impact wetland hydrology, leading to the desiccation of riparian zones and changes in species composition.

The primary cause of these changes is the direct diversion of water through canal networks, reducing downstream discharge even though the impoundment ratio (IR) index suggested minimal impact. This paradox underscores the necessity of considering actual water management practices rather than relying solely on hydrological indices. Additionally, seasonal shifts in flow availability, particularly an 18% reduction in base flow, exacerbate dry season water scarcity, affecting both aquatic habitats and dependent terrestrial ecosystems (Batalla et al. 2004; Chaudhari and Pokhrel 2022). Ecologically, these alterations compromise wetland hydrology, leading to habitat fragmentation and decreased biodiversity. The attenuation of flood peaks minimizes sediment deposition, which is critical for maintaining wetland fertility and sustaining aquatic vegetation (Bergkamp et al. 2000; Zeiringer et al. 2018). Reduced flow variability disrupts the lifecycle of aquatic species that rely on seasonal flooding cues, potentially leading to population declines (Richter et al. 1997). Moreover, increased low-flow frequencies, particularly in pre-monsoon and post-monsoon months, enhance the risk of hypoxic conditions, further stressing wetland ecology (Ward and Stanford 1979).

Overall, the results of this study support global evidence that damming leads to substantial hydrological and ecological transformations. The Atreyee River exhibits trends similar to other dammed systems, suggesting the need for adaptive water management strategies, such as controlled reservoir releases and sediment augmentation programs, to mitigate adverse impacts on wetland hydrology and biodiversity. Future research should integrate artificial intelligence-based hydrological modeling with ecological assessments to develop sustainable river basin management frameworks.

References

Alexandersson H, Moberg A (1997) Homogenization of Swedish temperature data. Part I: homogeneity test for linear trends. Int J Climatol J R Meteorol Soc 17(1):25–34

Batalla RJ, Gomez CM, Kondolf GM (2004) River impoundment and changes in flow regime, Ebro River basin, northeastern Spain. J Hydrol 290:117–136

Benn PC, Erskine WD (1994) Complex channel response to flow regulation: Cudgegong River below Windamere Dam, Australia. Appl Geogr 14(2):153–168

Bergkamp G, McCartney M, Dugan P, McNeely J, Acreman M (2000) Dams, ecosystem functions and environmental restoration. Themat Rev II 1:1–187

Billman GE (2012) Homeostasis: the dynamic self-regulatory process that maintains health and buffers against disease. In: Handbook of systems and complexity in health. Springer New York, New York, NY, pp 159–170

Breiman L (2001) Random forests. Mach Learn 45:5–32

Chaudhari S, Pokhrel Y (2022) Alteration of river flow and flood dynamics by existing and planned hydropower dams in the Amazon River Basin. Water Resour Res 58(5):e2021WR030555

Cuddy JD, Della Valle PA (1978) Measuring the instability of time series data. Oxf Bull Econ Stat 40(1)

Dyer F, ElSawah S, Croke B, Griffiths R, Harrison E, Lucena-Moya P, Jakeman A (2014) The effects of climate change on ecologically-relevant flow regime and water quality attributes. Stoch Env Res Risk Assess 28:67–82

Francke T, López Tarazón JA, Schröder B (2008) Estimation of suspended sediment concentration and yield using linear models, random forests and quantile regression forests. Hydrol Process Int J 22(25):4892–4904

Gao P, Mu XM, Wang F, Li R (2011) Changes in streamflow and sediment discharge and the response to human activities in the middle reaches of the Yellow River. Hydrol Earth Syst Sci 15(1):1–10

Graf WL (2006) Downstream hydrologic and geomorphic effects of large dams on American rivers. Geomorphology 79(3–4):336–360

Guo H, Hu Q, Zhang Q, Feng S (2012) Effects of the three gorges dam on Yangtze river flow and river interaction with Poyang Lake, China: 2003–2008. J Hydrol 416:19–27

Helsel DR, Hirsch RM (1993) Statistical methods in water resources, vol 49. Elsevier

Jenkins GM (1968) Spectral analysis and its applications

Kendall DG (1975) Rank correlation methods. Charles Griffin, London

Khatun R, Talukdar S, Pal S, Saha TK, Mahato S, Debanshi S, Mandal I (2021) Integrating remote sensing with swarm intelligence and artificial intelligence for modelling wetland habitat vulnerability in pursuance of damming. Eco Inform 64:101349

Kondolf GM, Batalla RJ (2005) Hydrological effects of dams and water diversions on rivers of Mediterranean-climate regions: examples from California. In: Developments in earth surface processes, vol 7. Elsevier, pp 197–211

Lehmann R, Totsche KU (2020) Multi-directional flow dynamics shape groundwater quality in sloping bedrock strata. J Hydrol 580:124291

Leopold LB, Wolman MG, Miller JP, Wohl EE (2020) Fluvial processes in geomorphology. Courier Dover Publications

Linsley Jr RK, Kohler MA, Paulhus JL (1975) Hydrology for engineers

Liro M (2014) Conceptual model for assessing the channel changes upstream from dam reservoir. Quaest Geogr 33(1):61–75

Mann HB (1945) Nonparametric tests against trend. Econ J Econ Soc 13(3):245–259

Monk WA, Wood PJ, Hannah DM, Wilson DA (2007) Selection of river flow indices for the assessment of hydroecological change. River Res Appl 23(1):113–122

Palmer M, Ruhi A (2019) Linkages between flow regime, biota, and ecosystem processes: implications for river restoration. Science 365(6459):eaaw2087

Pettitt AN (1979) Testing for bivariate normality using the empirical distribution function. Commun Stat Theory Methods 8(7):699–712

Petts GE (1984) Impounded rivers: perspectives for ecological management. Wiley, Chichester, p 326

Poff NL, Allan JD, Bain MB, Karr JR, Prestegaard KL, Richter BD, Sparks RE, Stromberg JC (1997) The natural flow regime: a paradigm for river conservation and restoration. Bioscience 47(11):769–784

Polikar R (2012) Ensemble learning. In: Ensemble machine learning: methods and applications, pp 1–34

Pourghasemi HR, Razavi-Termeh SV, Kariminejad N, Hong H, Chen W (2020) An assessment of metaheuristic approaches for flood assessment. J Hydrol 582:124536

Richter B, Baumgartner J, Wigington R, Braun D (1997) How much water does a river need? Freshw Biol 37(1):231–249

Schaefer D, Domroes M (2009) Recent climate change in Japan–spatial and temporal characteristics of trends of temperature. Clim past 5(1):13–19

Sen PK (1968) Estimates of the regression coefficient based on Kendall's tau. J Am Stat Assoc 63(324):1379–1389

Smakhtin VU, Shilpakar RL, Hughes DA (2006) Hydrology-based assessment of environmental flows: an example from Nepal. Hydrol Sci J 51(2):207–222

Tangang FT, Juneng L, Ahmad S (2007) Trend and interannual variability of temperature in Malaysia: 1961–2002. Theoret Appl Climatol 89:127–141

Von Storch H, Navarra A (eds) (1999) Analysis of climate variability: applications of statistical techniques. Springer Science & Business Media

Ward JV, Stanford JA (1979) Ecological factors controlling stream zoobenthos with emphasis on thermal modification of regulated streams. The ecology of regulated streams. Springer, Boston, MA, pp 35–55

Ward JV, Stanford JA (1995) Ecological connectivity in alluvial river ecosystems and its disruption by flow regulation. Regul Rivers Res Manage 11(1):105–119

Williams GP (1978) Bank full discharge of rivers. Water Resour Res 14(6):1141–1154

Yue S, Wang CY (2002) Applicability of prewhitening to eliminate the influence of serial correlation on the Mann Kendall test. Water Resour Res 38(6):1–4

Yue S, Pilon P, Cavadias G (2002) Power of the Mann-Kendall and Spearman's rho tests for detecting monotonic trends in hydrological series. J Hydrol 259(1–4):254–271

Zare M, Pourghasemi HR, Vafakhah M, Pradhan B (2013) Landslide susceptibility mapping at Vaz Watershed (Iran) using an artificial neural network model: a comparison between multilayer perceptron (MLP) and radial basic function (RBF) algorithms. Arab J Geosci 6:2873–2888

Zeiringer B, Seliger C, Greimel F, Schmutz S (2018) River hydrology, flow alteration, and environmental flow. In: Riverine ecosystem management: science for governing towards a sustainable future, pp 67–89

Zhang S, Lu XX (2009) Hydrological responses to precipitation variation and diverse human activities in a mountainous tributary of the lower Xijiang, China. CATENA 77(2):130–142

Zhang H, Liang Z, Liu H, Wang R, Liu Y (2020) Ensemble framework by using nature inspired algorithms for the early-stage forest fire rescue—a case study of dynamic optimization problems. Eng Appl Artif Intell 90:103517

3 Eco-hydrological Flow

Abstract

The indicator of hydrological alteration (IHA) and range of variability centric eco-hydrological flow in river assessment were tried to be done in this chapter. Ecological flow was also estimated in order to provide a guide toward restoration of river flow. Result shoed that mean monthly RVA-based flow failure rate ranges from 7.69 to 69.23%. This degree of hydrological alteration for different IHA groups range from − 0.65 to − 0.81 which signifies high degree of flow alteration. Integrated degree of impact due to hydrological control is 0.406 which also signifies high degree of flow alteration impact. Flow duration curve (FDC) analysis revealed that in eco-deficit is more often than eco-surplus. Estimation of environmental flow using global flow calculator revealed that the current flow state falls under largely modified environmental management classes (EMCs), and in order to restore flow, the amount of flow to be further released from dam was also computed. The chapter further stated that such change in eco-flow is harmful and going to be further harmful to the river ecology.

Keywords

Indicator of hydrological alteration · Flow failure · Degree of hydrological alteration · Environmental flow · Flow duration curve · Environmental management

3.1 Concept and Needs of Eco-hydrological Flow

Wassen and Grootjans (1996) defined eco-hydrology purely in reference to the processes and function of wetland. It is an interdisciplinary study of hydrology and associated ecology. It explains how hydrological factors determine the natural development of wet ecosystem, specifically in response to the functional values toward nature protection and restoration. It is the science of eco-hydrological aspects of ecology. Although the term was primarily used for wetland, now it is used in broad connotation. For instance, this term is also frequently used in river. In the case of river, this term refers to explore the specific flow condition that is needed to continue the existing ecosystem, its functions, and services (Schiemer 2016; Rai and Fulekar 2023). In order to tame river flow, different hydro-engineering structures like dam, barrage, embankment, canal system, etc., have been constructed and these have posed adverse impact on downstream flow (Kondolf and Yi 2022). In Chap. 2, it has been discussed how flow was changed in relation to damming. With this change, some flow-dependent ecological settings also changes significantly. For evaluating the flow change, environmental flow assessment (EFA) has become essential. EFA is an approach to assess how much flow should be allowed downstream and onto its floodplain after construction of damming in order to maintain valued features of the ecosystem (King et al. 2000). King

S. Pal et al., *Damming and Wetland Eco-hydrology*, GIScience and Geo-environmental Modelling,
https://doi.org/10.1007/978-3-032-11729-8_3

Table 3.1 IHA groups and their associated indicators adopted from Richter et al. (1996)

IHA group	Flow regime characteristic	Indicators of hydrologic alteration (IHA)
1. Magnitude of monthly water conditions	Magnitude/timing	IHA1–IHA12: Monthly mean values for each calendar month
2. Magnitude and duration of annual extremes	Magnitude/duration	IHA13–IHA24: 1–90 day annual minima/maxima, zero-flow days, base flow index
3. Timing of annual extremes	Timing	IHA25–IHA26: Julian dates of annual 1-day min and max
4. Frequency and duration of flow pulses	Frequency/duration	IHA27–IHA30: Number and mean duration of high and low pulses
5. Rate and frequency of changes	Rate of change	IHA31–IHA33: Rise rate, fall rate, and number of reversals

et al. (2000) rightly pointed extrapolated eco-hydrology both in reference to river and wetland. More than 200 definitions focusing low to high-confidence approaches are available in this regard. But their basic tune is same (Tharme 2003). Some frequently used approaches for eco-hydrological assessment based on flow matrices are range of variability approach (RVA), flow duration curve (FDC) analysis, indicator of hydrological alteration (IHA), assessment of flow failure rate, eco-hydro-deficit/surplus analysis, etc. In most of the cases, time series discharge or water level data is frequently used.

3.2 Method of Measuring Eco-hydrological Alteration

3.2.1 Indicator of Hydrological Alteration

The indicator of hydrological alteration (IHA), initially introduced by Richter et al. (1996), is a widely acknowledged framework for quantifying hydrological changes in rivers subjected to human-induced interventions or climate shifts. This tool encompasses 67 specific indicators, out of which 33 are commonly used to evaluate hydrologic changes, and the remaining 34 focus on ecological flow components. These indicators fall into five principal categories: (a) magnitude, referring to the volume or rate of discharge; (b) frequency, indicating how often specific flows occur; (c) duration, representing how long flow events persist; (d) timing, which relates to the regularity and seasonal position of events; and (e) rate of change, describing the speed at which water conditions shift (Richter et al. 1996). These aspects serve as the foundation for analyzing both hydrologic regimes and the related ecological implications—particularly when major changes such as dam construction alter the natural flow. A breakdown of IHA categories and their associated indicators is presented in Table 3.1.

3.2.2 Ecological Flow and Flow Failure Rate Computation

Ecological flow refers to the minimum water quantity needed to maintain ecosystem health within a river system. Assessing ecological flow before implementing flow-regulating structures like dams helps ensure sustainable design. Common techniques include using statistical percentiles (e.g., 20th–80th, 25th–75th), one or two standard deviations (SD), and the range of variability approach (RVA) proposed by Richter et al. (1997). These methods have been applied successfully in numerous studies (Laub et al. 2012; Pal and Talukdar 2020; Sarda and Pal 2022; Zheng et al. 2021). For this study, Mean $\pm$ 1 SD was used to establish ecological thresholds. Flow values within this range were considered to represent natural conditions; flows above it were categorized as ecologically surplus, and those below it as ecologically deficient. To quantify the flow failure rate (FFR), observed flows—monthly or yearly—were compared against RVA thresholds. The proportion of flows that fell outside this natural range constituted the FFR, evaluated for both pre- and post-dam conditions (Eq. 3.1).

$$\text{FFR} = \left(\frac{F_{\text{RVA}}}{T_{\text{FO}}}\right) \times 100\% \quad (3.1)$$

where FRVA = number of flow incidents beyond thresholds; TFO = total number of flow observations.

3.2.3 Measuring the Eco-hydrological Condition

The flow duration curve (FDC) is a valuable tool for analyzing shifts in hydrological regimes and their ecological implications. It plots discharge values against exceedance probabilities based on historical records. By comparing FDCs from periods before and after dam construction, one can visually interpret how often certain flows are met or exceeded. Following the approach of Vogel et al. (2007), this method allows for computing eco-deficit and eco-surplus—key indicators of hydrological health (Langat et al. 2019). If the post-dam FDC lies below the pre-dam FDC, it indicates eco-deficit; the opposite implies eco-surplus. The magnitude of the vertical gap between the two FDCs correlates with the degree of ecological stress or improvement. This method is broadly used in water resources engineering and planning (Pal et al. 2019; Burgan and Aksoy 2022). Due to the complexity of interpreting hydrographs alone, especially under dual influences of climate change and regulation, FDCs offer a more practical solution (Girihagama et al. 2022).

3.2.4 Measuring the Degree of Impact

To express the cumulative hydrological impact, Euclidean distance between pre- and post-regulation values of IHA indicators can be calculated and aggregated using Eq. 3.2. A larger distance reflects more significant deviation from natural conditions.

$$\text{ImHA} = \sqrt{\left[\frac{1}{n}\sum_{n}(I_i)^2\forall_j\right]} \quad (3.2)$$

where indicator of hydrological alteration (*ImHA*) is the overall degree of impact *j*, *i* refers to the percentage change of *i*th IHA indicator, and *n* shows to the total number of IHA taken into consideration. ImHA value ranges from 0 to 1. A value near 0 implies negligible hydrological change, whereas a value close to or exceeding 1 indicates severe impact.

3.2.5 Integrated Degree of Impact Due to Hydrological Alteration

In this chapter, alongside conventional assessment techniques, a novel methodology is introduced to evaluate the integrated indicator of hydrological alteration (IImHA). This composite index incorporates four critical components: (1) flow instability (IX) as detailed in Chap. 2, (2) flow failure rate (FFR), (3) the hydrological alteration index (ImHA) derived from IHA metrics, and (4) the net eco-hydrological deficit/surplus (NED/S). The procedures for calculating IX, FFR, and ImHA were discussed in earlier sections.

To standardize the evaluation, both IX and FFR were normalized using the method provided in Eq. 3.3. The NED/S component was derived by calculating the proportion of the gap between the regulated and unregulated flow duration curves (FDCs) to the total area under the unregulated FDC, as described in Eq. 3.4. After normalizing all indicators, a simple arithmetic average (without applying weights) was taken to compute the final IImHA score using Eq. 3.5. This integrated metric ranges between 0 and 1, where a value of 0 indicates no detectable alteration from natural flow conditions, and a value approaching 1 reflects a significantly impacted or highly altered hydrological regime.

$$X_{\text{Nor}} = \frac{X - X_{\text{Min}}}{X_{\text{Max}} - X_{\text{Min}}} \quad (3.3)$$

where X_{Nor} = normalized observation, X_{Max} = maximum observation, X = actual observation, X_{Min} = minimum observation.

$$\text{NED}/S = (\Delta\text{FDC}/\text{FDC}_N) \quad (3.4)$$

where NED/S = net eco-deficit/surplus, ΔFDC = gap between regulated and unregulated FDCs, FDC_N = area under unregulated FDC.

$$\text{IImHA} = \frac{(\text{CV} + \text{FFR} + \text{ImHA} + \text{NED}/S)}{N} \quad (3.5)$$

where IImHA = integrated degree of impact of hydrological alteration; N = total number of indices considered (here four indices taken).

3.2.6 Estimation of Environmental Flow

Smakhtin and Anputhas (2006) developed the Global Environmental Flow Calculator (GEFC) as a tool to estimate the environmental flow requirements of river systems using long-term hydrological records. This model operates on a set of seventeen predefined percentiles—ranging from extreme low to high flows (0.01, 0.1, 1, 5, 10, 20, 30, 40, 50, 60, 70, 80, 90, 95, 99, 99.9, and 99.99%)—to comprehensively represent flow variability across the entire spectrum. Within this framework, rivers are categorized into six Ecological Management Classes (EMCs)—A through F—based on the degree of deviation from their natural hydrological regime. These classes provide a comparative evaluation of the ecological condition of river flows. Classes A and B represent largely undisturbed or natural flow conditions, while Classes E and F denote highly altered and ecologically compromised states. Class D is regarded as the lowest acceptable management standard, and Class C reflects a condition that, though altered, remains ecologically sustainable. An overview of these classes is presented in Table 3.2. To begin the estimation, a flow duration curve (FDC) is generated using the 17 fixed percentiles to serve as a baseline or reference curve. The environmental flow thresholds for each EMC are then derived by systematically adjusting this reference curve downward by fixed percentages corresponding to each class. The

Table 3.2 Environmental flow management classes and their characteristics and management perspectives

Class	Characteristics	Management perspectives
A (natural flow regime)	Minimal or no human alterations; ecosystem maintains natural variability	Preserve and protect natural hydrological conditions through conservation efforts and stringent water use regulations
B (slightly modified)	Small deviations from natural flow patterns; minor ecological impacts	Implement sustainable water use practices, minor flow adjustments, and habitat conservation programs
C (moderately modified)	Noticeable changes in flow regime; some species and ecological functions affected	Maintain ecological balance with controlled water releases, seasonal flow management, and habitat restoration
D (largely modified)	Significant alteration of flow; substantial ecological changes; some species decline	Implement restoration programs, adjust dam operations, and enhance environmental flow allocations
E (seriously modified)	Major loss of ecosystem functions; high stress on aquatic species; frequent low flows	Rehabilitate critical habitats, introduce compensatory mitigation measures, and improve water governance
F (critically modified)	Extremely altered system; severe ecological degradation; loss of key species	Prioritize ecosystem recovery, enforce strict water regulations, and explore alternative water sources

environmental flow requirement for each class is calculated by dividing the cumulative flow values at these percentiles by the Mean Annual Flow (MAF), which yields a percentage reflecting the environmental water demand of that class. In practical application, the post-dam FDC is superimposed onto the estimated FDCs for each EMC. This allows for both visual and analytical identification of the ecological class into which the current regulated flow falls. Such comparison provides important insights into the ecological integrity and long-term flow sustainability of the river under altered hydrological conditions.

3.2.7 Range of Variability Approach

The detailed procedure for applying the RVA has already been outlined earlier in Chap. 2. This method provides a robust framework for evaluating hydrological changes by comparing flow statistics across different temporal regimes.

3.3 Results

3.3.1 Eco-hydrological Thresholds and Flow Failure Rates

Monthly river discharge was categorized using eco-hydrological thresholds into three distinct classes:

1. **Ecological flow optima**, where flows fall within the interquartile range (25th to 75th percentiles),
2. **flow failures above the upper threshold** (exceeding the 75th percentile), and
3. **flow failures below the lower threshold** (falling below the 25th percentile).

Figure 3.1 illustrates the temporal distribution of monthly flows from 1993 to 2024, showing how individual data points align with or deviate from the calculated threshold bands. This visualization highlights the distribution of flows relative to ecological norms and helps in detecting conditions of ecological stress. A noticeable increase in flow failure events is evident during the post-dam period, suggesting an intensification of hydrological alteration. When failure rates were analyzed seasonally, both the pre-monsoon and post-monsoon months exhibited consistently high rates of deviation—often surpassing 45%, as shown in Table 3.3. Further analysis revealed that failures below the lower threshold were far more common than exceedances above the upper limit (Fig. 3.1). These deficits are largely attributed to reduced rainfall and intensive flow diversion associated with dam operations. In contrast, the pre-dam period recorded very few instances where flow dropped below the lower ecological limit. When comparing maximum and minimum monthly flows with their corresponding threshold bands, the findings mirrored those based on mean flow analysis, reaffirming the trend of increased hydrological stress and ecological vulnerability during the regulated flow period.

3.3.2 Degree of Impact from Hydrological Alteration

The extent of hydrological modification was evaluated using group-specific IHA values, with results presented in Table 3.4. The degree of alteration across different IHA categories was found to range between − 0.65 and − 0.81, indicating a substantial departure from natural flow conditions. Specifically, the computed values associated with various hydrological components were as follows: magnitude (− 0.66), duration (− 0.68), timing (− 0.72), frequency of high- and low-flow pulses (− 0.81), and rate of change (− 0.65). These results collectively point toward a significant level of eco-hydrological disruption. Such pronounced changes are likely to lead to increasing ecological pressure, potentially threatening the sustainability of aquatic ecosystems and altering the river's natural hydrological dynamics.

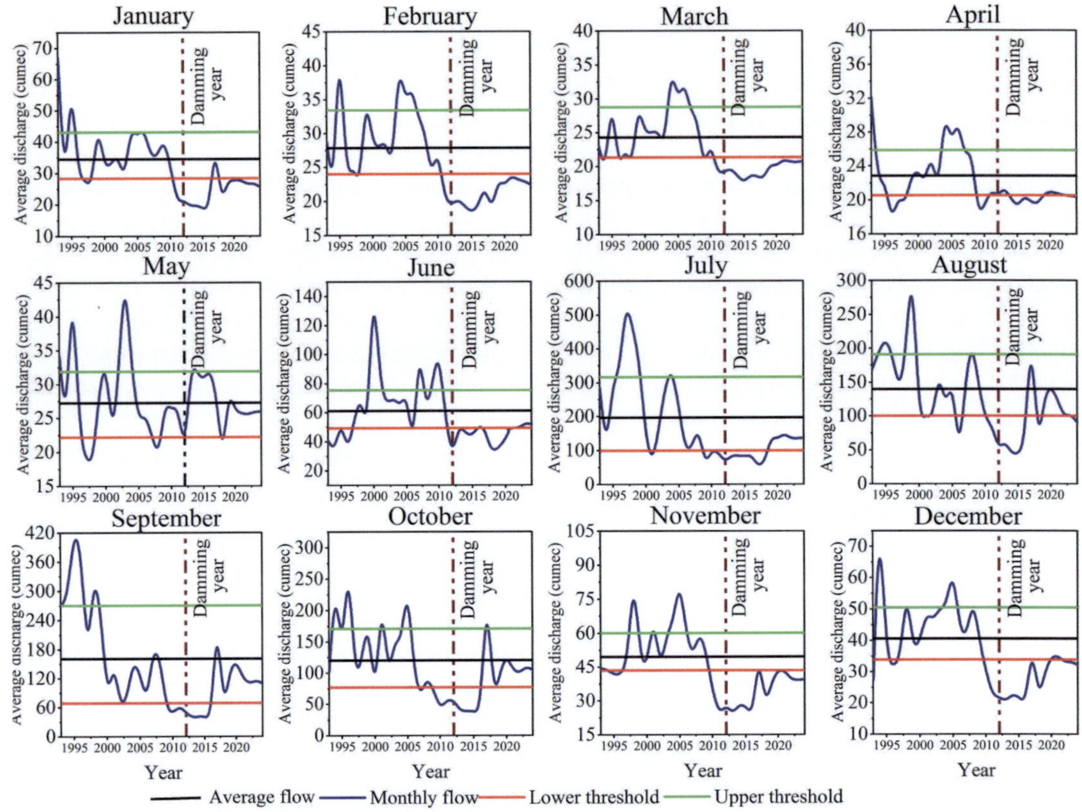

Fig. 3.1 displays the monthly discharge pattern of the Atreyee River relative to its upper and lower RVA thresholds, helping to identify ecological flow conditions and deviations

Table 3.3 Monthly average discharges in reference to pre- and post-dam periods; computed RVA thresholds, failure rates, and degree of hydrological alteration

Month	Mean discharge (m^3/s)		RVA thresholds (m^3/s)		Failure rate (%)
	Before dam	After dam	Lower limit	Higher limit	
January	37.20	24.68	24.89	49.50	46.15
February	29.42	21.15	22.47	36.37	69.23
March	25.11	19.52	20.23	29.99	61.54
April	23.40	20.37	19.21	27.58	7.69
May	28.01	27.15	19.57	36.44	15.38
June	67.48	46.04	36.47	98.49	15.38
July	223.42	103.72	79.87	366.96	23.08
August	151.80	94.96	84.46	219.14	53.85
September	177.53	96.67	57.16	297.90	46.15
October	131.87	88.81	56.10	207.65	46.15
November	52.64	34.84	36.63	68.65	46.15
December	43.90	28.08	29.49	58.31	46.15

Table 3.4 Mean degree of hydrological alteration (HA) of five IHA groups

IHA group/parameters	Group 1	Group 2	Group 3	Group 4	Group 5
Hydrological alteration	– 0.66	– 0.68	– 0.72	– 0.81	– 0.65
Remarks	High HA	High HA	High HA	High HA	High HA

Note Degree of hydrological alteration as noted in Remark part is based on Richter et al. (1997) recommended classes

3.3.3 Eco-hydrological State of the River

Figure 3.2 presents monthly FDCs comparing unregulated (pre-dam) and regulated (post-dam) discharge data to assess ecological surplus and deficit conditions. Across all months, the FDC corresponding to the natural flow regime consistently lies above that of the regulated condition, indicating a clear eco-hydrological deficit during the post-dam period. However, the intensity of this deficit varied from month to month. Notably, the monsoon and post-monsoon months exhibited the most significant differences, as shown in Fig. 3.2. For example, in July, under natural flow conditions, there was nearly a 100% probability of the river delivering 300 m^3/s, whereas under regulated conditions, discharge remained below 100 m^3/s at all exceedance probabilities. The figure illustrates that each month has experienced some level of eco-deficit, reflecting an increasing trend toward eco-hydrological stress. Such conditions often result in the loss of aquatic species, degradation of biodiversity, and deterioration of habitats critical for spawning and breeding. These findings align with previous research; for instance, Gain and Giupponi (2014) reported similar ecological consequences due to altered flow regimes. Studies on China's Yangtze River also confirm this trend, with Wang et al. (2016) documenting eco-deficit-driven impacts. Likewise, Ren et al. (2019) found evidence of ecological degradation in the Wei River as a result of reduced flow volumes. The negative effects of hydrological regulation on riverine ecosystems were also observed in India by Kumar and Jayakumar (2020). Furthermore, Aziz et al. (2021) highlighted that reduced water availability following damming led to the local extinction of several key species in the Hakaluki Haor wetland of northeastern Bangladesh.

Total seven river sites were surveyed in order to explore different hydrological and ecological transformations of Atreyee River. Qualitative answers and some quantitative information regarding this are presented in Fig. 3.3. Flow reduction, flood water depth, frequency, and flood extent were reported highly affected by damming incident. All sites witnessed a decreasing trend. This result also supports the eco-hydrological change as reported in Fig. 3.2.

3.3.4 Integrated Degree of Impact from Hydrological Regulation

The IImHA, calculated using a suite of eco-hydrological metrics, revealed a moderate level of ecological impact resulting from flow regulation, with an overall IImHA value of 0.406. The component indicators contributing to this value included flow variability (0.527), flow failure rate (FFR) (0.217), ImHA based on IHA metrics (0.526), and the net eco-deficit/surplus (NED/S) (0.355), as summarized in Table 3.5. These individual metrics reflect a range of conditions from low to moderate impact, with FFR being the only indicator suggesting a relatively low degree of disturbance. The results emphasize that, while the system has not reached critical levels of ecological disruption, the trajectory of increasing water abstraction poses a serious risk. Therefore, proactive management strategies are essential to prevent further deterioration and to promote long-term ecological resilience.

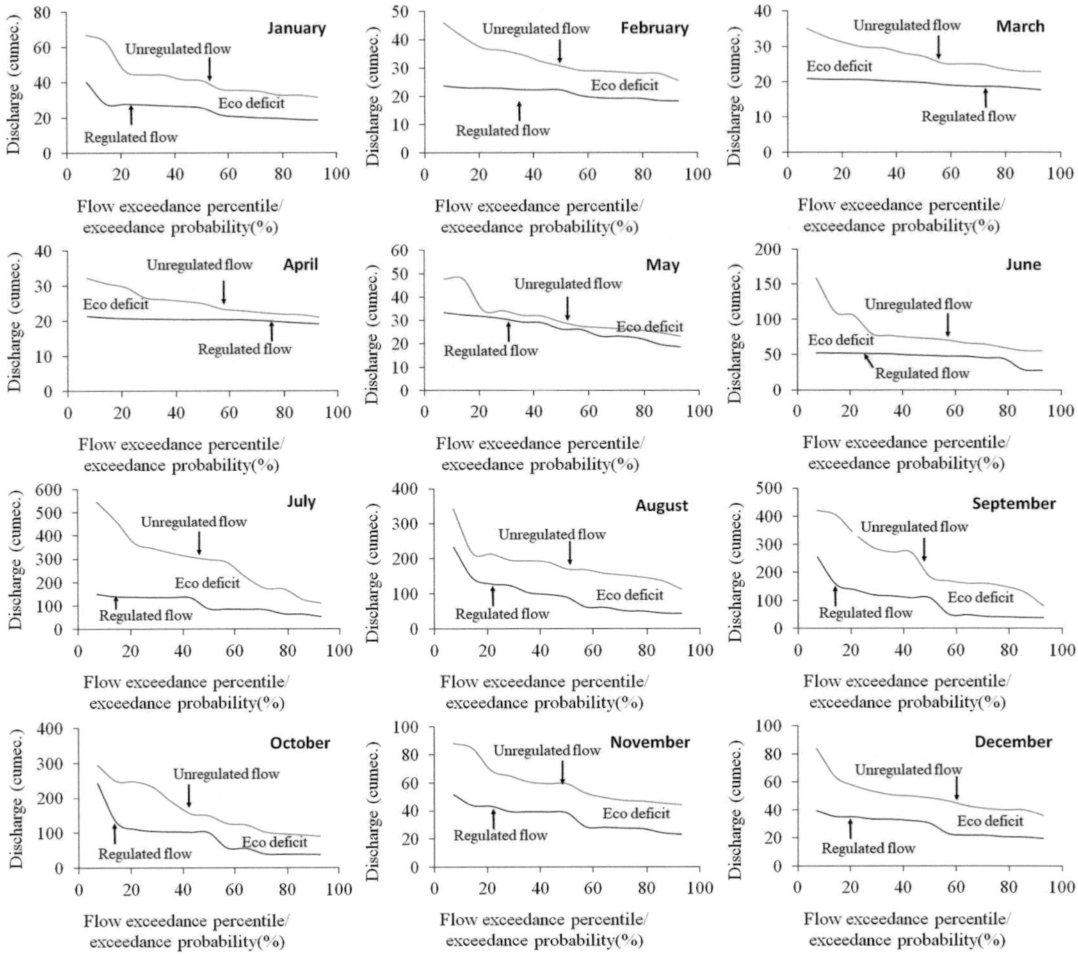

Fig. 3.2 Monthly FDCs for the Atreyee River comparing pre-dam (unregulated) and post-dam (regulated) conditions. The green line represents the natural flow regime, while the violet line indicates the regulated discharge pattern. The visible separation between the two curves illustrates a condition of **eco-hydrological deficit**, where flow availability has declined following regulation

3.3.5 Environmental Flow Requirement

Regulated and unregulated flow states in reference to calculated monthly EMC are shown in Fig. 3.4 in order to present the accordance or discordance with the EMC classes. This could clearly help to explore how the current flow is matched with which EMC class and what is the possible flow requirement of the river concerned to maintain the ecological balance. Pre- and post-dam flow references could help to compare the changing flow position within the warp of EMCs. This is very crucial for framing future strategies of river water resource management. As per the mechanism of EMC calculation, the ecological flow requirement is diverse for different EMCs and it is high in the case of higher EMC (Poff and Zimmerman 2010).

For the sake of human civilization, use of river water is not any new, but there should be a limit. According to Sood et al. (2017), Richter et al. (2012) flow amendment could be permitted up to moderately modified EMC. Largely, severely and critically modified EMCs are not ecologically sustainable; therefore, interpretation of the present result is done accordingly. Table 3.6 shows esti-

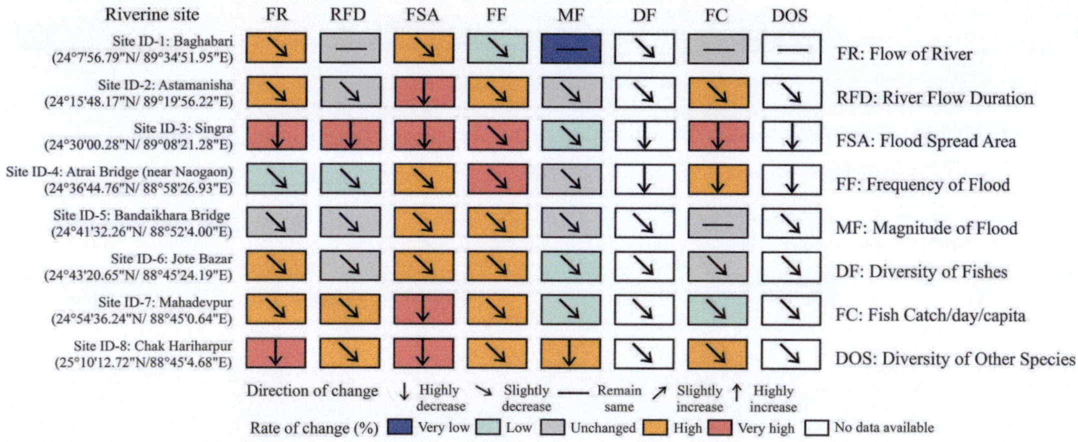

Fig. 3.3 Field survey-based hydrological and ecological transformation parameters as perceived by bank dwellers

Table 3.5 Integrated degree of impact of hydrological alteration

Flow instability (IX)	Flow failure rate (FFR)	Degree of the impact of hydrological alteration (ImHA)	Net Eco-deficit/ surplus (NED/S)	Integrated degree of impact of hydrological alteration (IImHA)
0.35	0.46	0.73	0.63	0.54

mated annual discharge at different exceedance probabilities. Figure 3.4 depicts that post-dam monthly flow condition in response to EMCs at different exceedance flow probabilities, and considerable variation is noted. Comparison of estimated annual discharge of different EMCs and post-dam flow revealed the fact that annual flow condition is matched with largely modified EMC.

3.4 Discussion

The findings of this study provide significant insights into the impact of hydrological modifications on the eco-hydrological regime of the Atreyee River. The results demonstrate that damming has led to considerable changes in flow dynamics, affecting both seasonal and long-term hydrological patterns. One of the most striking observations is the increase in flow failure rates, particularly during pre-monsoon and post-monsoon months. This suggests that water availability during these critical ecological periods has been substantially reduced, which can have severe consequences for aquatic biodiversity, habitat stability, and river-dependent livelihoods. Similar trends have been reported in other studies on regulated rivers, where damming has disrupted natural flow variability, leading to shifts in ecosystem functions and species distributions. This aligns with the findings of Gain and Giupponi (2014), who demonstrated that the Farakka Barrage on the Ganges River caused a decline in dry-season flows, significantly affecting aquatic habitat availability and ecosystem sustainability. Similarly, Ren et al. (2019) reported that flow regulation in the Wei River led to reduced base flows and an overall eco-hydrological deficit, making it difficult for dependent ecosystems to sustain their natural processes.

The flow duration curve (FDC) analysis further reinforces the presence of an eco-deficit state in the Atreyee River, as post-dam FDCs consistently remain below pre-dam FDCs across all months. This implies that the river no longer maintains its natural hydrological variability, which is essential for sustaining ecosystem functions such as sediment transport, fish migration, and floodplain nourishment. Similar findings have been observed in the Yangtze River, where Wang et al. (2016) found that dam-induced flow reduc-

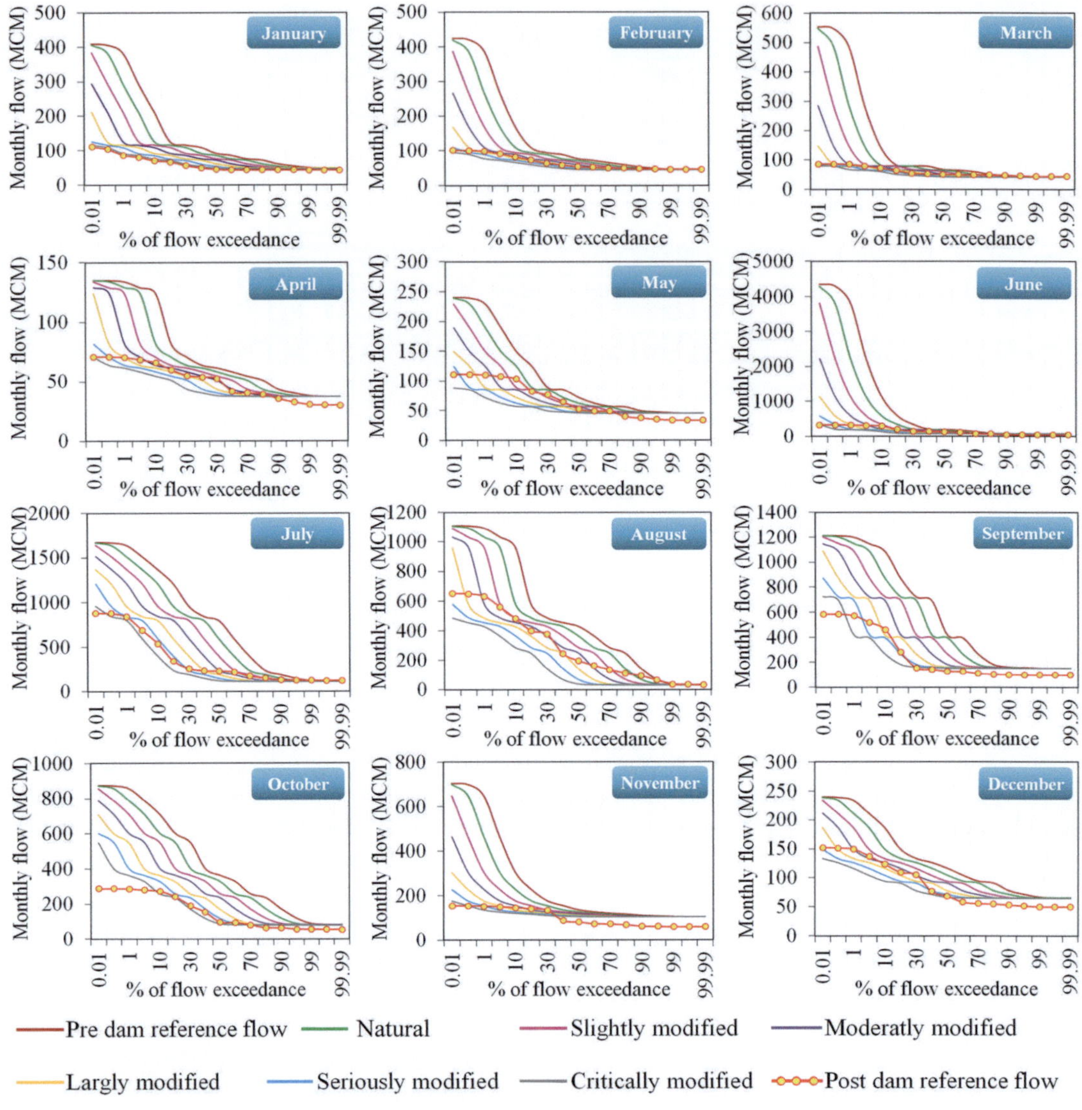

Fig. 3.4 Post-dam flow state in reference to computed monthly EMC

tions led to significant alterations in the availability of suitable spawning grounds for native fish species. Likewise, Kumar and Jayakumar (2020) studied hydrological alterations in the Krishna River Basin and found that regulated flow conditions contributed to an increased frequency of low-flow events, disrupting riparian ecosystems and fish migration patterns. The Atreyee River's increasing eco-deficit condition, as revealed in this study, suggests a pressing need for adopting environmental flow management strategies to restore the river's ecological balance.

The indicator of hydrological alteration (IHA) assessment provides additional insights into the magnitude of hydrological change. The results show that all five key hydrological parameters—magnitude, duration, timing, frequency, and rate of change—exhibit significant alterations, confirming that damming has caused substantial shifts in the river's eco-hydrological dynamics. The degree of hydrological alteration in the Atreyee River falls within the range of − 0.65 to − 0.81, indicating a high level of disruption. This is consistent with the findings of Richter et al.

Table 3.6 Yearly estimated average flow requirement of different EMC classes, reference pre- and post-dam flow probability

Flow exceedance (%)	Pre-dam reference flow	Natural	Slightly modified	Moderately modified	Largely modified	Seriously modified	Critically modified	Post-dam reference flow
0.01	1706	1652	1199	791	525	329	205	673
0.1	1652	1199	791	525	329	205	154	671
1	1199	791	525	329	205	154	118	652
5	791	525	329	205	154	118	92.8	366
10	525	329	205	154	118	92.8	79.3	316
20	329	205	154	118	92.8	79.3	67.5	191
30	205	154	118	92.8	79.3	67.5	56.9	132
40	154	118	92.8	79.3	67.5	56.9	51.4	105
50	118	92.8	79.3	67.5	56.9	51.4	47.7	87.5
60	92.8	79.3	67.5	56.9	51.4	47.7	43.4	70.8
70	79.3	67.5	56.9	51.4	47.7	43.4	43	55.5
80	67.5	56.9	51.4	47.7	43.4	43	42.6	53.2
90	56.9	51.4	47.7	43.4	43	42.6	42.2	50.1
95	51.4	47.7	43.4	43	42.6	42.2	41.8	48.5
99	47.7	43.4	43	42.6	42.2	41.8	41.4	44.5
99.9	43.4	43	42.6	42.2	41.8	41.4	41	43.9
99.99	43	42.6	42.2	41.8	41.4	41	40.6	43.8

(1996), who proposed that alterations beyond − 0.50 represent a severe impact on natural hydrological conditions, leading to major ecological consequences. Similar high degrees of hydrological alteration have been documented in other studies, such as the Colorado River Basin, where Poff and Zimmerman (2010) demonstrated that flow regulation led to changes in the timing of peak flows, affecting floodplain connectivity and sediment transport. The Atreyee River's hydrological alteration follows a similar trajectory, suggesting that urgent mitigation measures are needed to restore the natural flow variability that supports aquatic and riparian ecosystems.

The integrated degree of hydrological alteration index (IImHA) calculated in this study provides a comprehensive assessment of the overall eco-hydrological impact. The computed IImHA value of 0.406 indicates a moderate to high level of hydrological alteration, emphasizing the need for sustainable water management interventions. The individual indices contributing to IImHA, such as flow variability (0.527), flow failure rate (0.217), hydrological alteration degree (0.526), and eco-deficit (0.355), collectively highlight the severity of eco-hydrological stress. A similar approach was adopted by Aziz et al. (2021) in their study of Hakaluki Haor in Bangladesh, where declining water availability due to both climate change and anthropogenic factors led to the loss of key fish species and wetland biodiversity. Likewise, Sarda and Pal (2022) used the IHA approach to evaluate the Tangon River basin and found that hydrological alterations were responsible for increased habitat fragmentation and ecological stress. Given the growing impact of damming on river systems, this study's findings emphasize the need for adaptive management strategies to ensure that hydrological modifications do not irreversibly harm ecological functions.

The environmental flow assessment conducted in this study further highlights the ecological consequences of flow regulation. The flow duration curve (FDC) shifting method shows that post-dam flow conditions align with the "largely modified" ecological management class (EMC), suggesting that the river is experiencing substantial hydrological stress. Richter et al. (2012) proposed that flow alterations exceeding moderately modified conditions pose a significant risk to aquatic biodiversity and should be addressed through restoration strategies such as controlled flow releases and environmental flow allocations. Studies by Smakhtin and Anputhas (2006) demonstrated that maintaining flow conditions within the moderately modified EMC is necessary to support long-term ecological sustainability. The findings from the Atreyee River reinforce these conclusions, indicating that failure to address environmental flow requirements could lead to long-term ecological degradation.

The results of this study contribute to the broader field of eco-hydrology and river basin management, offering empirical evidence of how damming and flow regulation alter hydrological and ecological systems. By integrating multiple assessment approaches-including IHA, FDC analysis, and environmental flow assessments—this research provides a holistic understanding of hydrological changes in a regulated river system. The study's findings underscore the need for—(i) controlled flow releases should be integrated into dam operations to maintain essential ecological functions. (ii) Sustainable water allocation policies should be designed to balance human water demand with ecological sustainability.

References

Aziz MSB, Hasan NA, Mondol MMR, Alam MM, Haque MM (2021) Decline in fish species diversity due to climatic and anthropogenic factors in Hakaluki Haor, an ecologically critical wetland in northeast Bangladesh. Heliyon 7(1)

Burgan HI, Aksoy H (2022) Daily flow duration curve model for ungauged intermittent subbasins of gauged rivers. J Hydrol 604:127249

Gain AK, Giupponi C (2014) Impact of the Farakka Dam on thresholds of the hydrologic flow regime in the Lower Ganges River Basin (Bangladesh). Water 6(8):2501–2518

Gao Y, Vogel RM, Kroll CN, Poff NL, Olden JD (2009) Development of representative indicators of hydrologic alteration. J Hydrol 374(1–2):136–147

Girihagama L, Naveed Khaliq M, Lamontagne P, Perdikaris J, Roy R, Sushama L, Elshorbagy A (2022) Streamflow modelling and forecasting for Canadian watersheds using LSTM networks with attention mechanism. Neural Comput Appl 34(22):19995–20015

King JM, Tharme RE, De Villiers MS (2000) Environmental flow assessments for rivers: manual for the Building Block Methodology. Water Research Commission, Pretoria, p 340

Kondolf M, Yi J (2022) Dam renovation to prolong reservoir life and mitigate dam impacts. Water 14(9):1464

Kumar AU, Jayakumar KV (2020) Hydrological alterations due to anthropogenic activities in Krishna River Basin, India. Ecol Indic 108:105663

Langat PK, Kumar L, Koech R (2019) Monitoring river channel dynamics using remote sensing and GIS techniques. Geomorphology 325:92–102

Laub BG, Baker DW, Bledsoe BP, Palmer MA (2012) Range of variability of channel complexity in urban, restored and forested reference streams. Freshw Biol 57(5):1076–1095

Pal S, Talukdar S (2020) Assessing the role of hydrological modifications on land use/land cover dynamics in Punarbhaba river basin of Indo-Bangladesh. Environ Dev Sustain 22:363–382

Pal S, Saha A, Das T (2019) Analysis of flow modifications and stress in the Tangon river basin of the Barind tract. Int J River Basin Manag 17(3):301–321

Poff NL, Zimmerman JK (2010) Ecological responses to altered flow regimes: a literature review to inform the science and management of environmental flows. Freshw Biol 55(1):194–205

Rai A, Fulekar MH (2023) Eco-hydrology: conservation of water resources. In: Climate change and sustainable development. CRC Press, pp 171–179

Ren K, Huang S, Huang Q, Wang H, Leng G, Cheng L et al (2019) A nature-based reservoir optimization model for resolving the conflict in human water demand and riverine ecosystem protection. J Clean Prod 231:406–418

Richter BD, Baumgartner JV, Powell J, Braun DP (1996) A method for assessing hydrologic alteration within ecosystems. Conserv Biol 10(4):1163–1174

Richter B, Baumgartner J, Wigington R, Braun D (1997) How much water does a river need? Freshw Biol 37(1):231–249

Richter BD, Davis MM, Apse C, Konrad C (2012) A presumptive standard for environmental flow protection. River Res Appl 28(8):1312–1321

Sarda R, Pal S (2022) Evaluating damming effect on eco-hydrological alteration in river and wetland using indicators of hydrological alteration. Geocarto Int 37(27):16132–16156

Schiemer F (2016) Building an eco-hydrological framework for the management of large river systems. Ecohydrol Hydrobiol 16(1):19–25

Smakhtin VU, Anputhas M (2006) An assessment of EF requirements of Indian river basins. Research

Report 107, International Water Management Institute, Colombo, Sri Lanka

Sood A, Smakhtin V, Eriyagama N, Villholth KG, Liyanage N, Wada Y et al (2017) Global environmental flow information for the sustainable development goals, vol 168. International Water Management Institute (IWMI)

Tharme RE (2003) A global perspective on environmental flow assessment: emerging trends in the development and application of environmental flow methodologies for rivers. River Res Appl 19(5–6):397–441

Vogel RM, Sieber J, Archfield SA, Smith MP, Apse CD, Huber-Lee A (2007) Relations among storage, yield, and instream flow. Water Resour Res 43(5)

Wang Y, Rhoads BL, Wang D (2016) Assessment of the flow regime alterations in the middle reach of the Yangtze River associated with dam construction: potential ecological implications. Hydrol Process 30(21):3949–3966

Wassen MJ, Grootjans AP (1996) Ecohydrology: an interdisciplinary approach for wetland management and restoration. Vegetatio 126(1):1–4

Zheng X, Yang T, Cui T, Xu C, Zhou X, Li Z et al (2021) A revised range of variability approach considering the morphological alteration of hydrological indicators. Stoch Env Res Risk Assess 35:1783–1803

4 Floodplain Wetland Mapping, Simulation, and Prediction

Abstract

This chapter explored different approaches of wetland mapping, monitoring, and area prediction. Spectral water indices, machine learning (ML) classifier, parametric image classification, and field parameters-based approaches were adopted for this. Artificial neural network (ANN)-cellular automata (CA) was applied for areal prediction of wetland area. Wetland maps were produced using different approaches. Normalized Difference Water Index (NDWI) predicted that in post-monsoon 2000 (pre-dam), wetland area was 1680 km^2 which was reduced to 802 km^2 in 2013 (post-dam) and continued to decline over progress of time. Wetland area prediction also assumed the likelihood of declining wetland area. Very high wetland extent during monsoon and post-monsoon seasons and low during pre-monsoon season is concomitant with seasonal rainfall pattern of this region. However, growing wetland loss is an ensuing ecological threat and needs due attention.

Keywords

Wetland mapping · Spectral water indices · Machine learning (ML) · Image classification · Area prediction · Cellular automata (CA)

4.1 Concept of Floodplain Wetland

The floodplain wetland is defined as the low-lying areas along the rivers, which are seasonally flooded regularly due to overspilling from the main river and its tributary system. India has a good numbers of floodplain wetlands, especially in the Ganga and Brahmaputra river basins (covering 0.35 million hectares areas) (Handbook of Fisheries and Aquaculture, 2011). These floodplain wetlands are of different types including oxbow lakes, meander, sloughs, small depression, palaeo channel, marsh, back swamps, etc. (Sinha and Tandon 2014; Islam 2016). These wetlands in India have different names in different regions, like *Beel* (Assam, West Bengal, Arunachal Pradesh, Tripura, and Meghalaya), *Pats* (Manipur), *Chaurs and Dhars* (Northern and South-Eastern West Bengal and Bihar). These wetlands are ecologically vibrant and economically remunerative. Rapid loss of wetland exposes both marginal peoples reliant on those wetlands. Scholars identified many reasons for wetland transformation like agricultural encroachment, built encroachment, fragmentation, drainage modification, groundwater lowering, and so on (Singh et al. 2022; Pal et al. 2022; Gupta et al. 2023; Ahmad et al. 2024). Singh

S. Pal et al., *Damming and Wetland Eco-hydrology*, GIScience and Geo-environmental Modelling,
https://doi.org/10.1007/978-3-032-11729-8_4

and Sinha (2022) rightly pointed out that lack of boundary of that wetland is one of the major causes behind growing insecurity and wetland loss. So, inventorying and mapping wetland is very essential.

Wetland delineation is a very indispensable job for the implementation of proper management strategies to diminish the quick pace of wetland transformation, particularly in floodplain regions (Debanshi and Pal 2020; Saha et al. 2021). Seasonally inundated floodplain wetlands are hydrologically dynamic due to significant variation of seasonal rainfall as well as flow volume of the concerned river. In pre-monsoon seasonal scarcity of rainfall exposes wider part of the wetland facilitating people to use that land for other purposes, particularly seasonal agricultural land. Rigorous such process gradually creates ambient environment for wetland transformation. Hence, these wetlands are highly susceptible to conversion. Areal dynamics of wetland is strongly associated with hydrological dynamics of the wetland. So, analysis of hydrological characters is also very necessary.

In subhumid region, wetland appearance in floodplain region is very irregular due to uncertain and arrhythmic pattern of seasonal rainfall. So, in a same season, wetland extent is highly variable. How frequently a wetland pixel receives water is also an issue of analysis. Length of hydro-duration in a year, water depth fluctuation, and seasonal range of water depth fluctuation are some relevant issues that need to be explain while dealing with hydrological character of the wetland identified. So, this chapter focuses on the approaches wetland mapping, its likelihood of change and changes in hydrological characters. All these are tried to be analyzed in reference to pre- and post-dam periods in order to present the changes that occurred.

4.2 Approaches of Delineation

Different approaches including (1) spectral water indices (2) land use/land cover classification (3) machine learning-based approach could be considered for wetland delineation. First two approaches were frequently used for this purpose.

4.2.1 Spectral Water Indices

Landsat-5 TM and Landsat-8 OLI satellite images obtained from the US Geological Survey (USGS) Global Visualization Viewer. The obtained Landsat data (Level 1 Terrain Corrected (L1T) product) were pre-georeferenced to UTM zone 45 North projection using WGS-84 datum. The detailed specification of the satellite imageries is mentioned in Table 4.1.

To accurately differentiate wetland features from surrounding land cover, a variety of spectral indices were applied. These include the Normalized Difference Vegetation Index (NDVI) (Jones and Vaughan 2010), Normalized Difference Water Index (NDWI) (McFeeters 1996), and Normalized Difference Moisture Index (NDMI) (Wilson and Sader 2002). Additional indices such as Modified NDWI (MNDWI) (Xu 2006), Water Ratio Index (WRI) (Shen and Li 2010), Normalized Difference Pond Index (NDPI) (Lacaux et al. 2007), Automated Water Extraction Index (AWEI) (Feyisa et al. 2014), and Re-modified NDWI (RmNDWI) (Debanshi and Pal 2020) were also utilized. These indices are well-established

Table 4.1 Satellite datasets employed for delineating wetland areas in the lower Atreyee River basin

Season	Period of acquisition year	Satellite	Sensor	Path/Row	Total Number of imageries	Spatial Resolution (m)	Note
Pre-monsoon (Mar–May)	1990–2024	Landsat-5	TM	138/43	11	30	Cloud cover less than 10%
		Landsat-8	OLI-TIRS		9		
Post-monsoon (Oct–Dec)		Landsat-5	TM		12		
		Landsat-8	OLI-TIRS		11		

tools for effectively distinguishing water bodies from non-water features in satellite imagery. All the approaches are not equally suitable for all the geographical regions and all these have their own merits and demerits. Along with merit of these indices, spectral nearness of objects, vegetation cover or any other cover over the wetland strongly hinders against the effective application of spectral water index (Ma et al. 2019).

$$\mathrm{NDVI} = \frac{(B_{\mathrm{NIR}} - B_{\mathrm{RED}})}{(B_{\mathrm{NIR}} + B_{\mathrm{RED}})} \tag{4.1}$$

$$\mathrm{NDWI} = \frac{(B_{\mathrm{GREEN}} - B_{\mathrm{NIR}})}{(B_{\mathrm{GREEN}} + B_{\mathrm{NIR}})} \tag{4.2}$$

$$\mathrm{NDMI} = \frac{(B_{\mathrm{NIR}} - B_{\mathrm{MIR}})}{(B_{\mathrm{NIR}} + B_{\mathrm{MIR}})} \tag{4.3}$$

$$\mathrm{MNDWI} = \frac{(B_{\mathrm{GREEN}} - B_{\mathrm{MIR}})}{(B_{\mathrm{GREEN}} + B_{\mathrm{MIR}})} \tag{4.4}$$

$$\mathrm{WRI} = \frac{(B_{\mathrm{GREEN}} + B_{\mathrm{RED}})}{(B_{\mathrm{NIR}} + B_{\mathrm{MIR}})} \tag{4.5}$$

$$\mathrm{NDPI} = \frac{(B_{\mathrm{MIR}} - B_{\mathrm{GREEN}})}{(B_{\mathrm{MIR}} + B_{\mathrm{GREEN}})} \tag{4.6}$$

$$\mathrm{AWEI} = B_{\mathrm{BLUE}} - 0.25 \times B_{\mathrm{GREEN}} - 1.03 \times B_{\mathrm{NIR}} - 0.8 \tag{4.7}$$

$$\mathrm{RmNDWI} = \frac{(B_{\mathrm{RED}} - B_{\mathrm{MIR}})}{(B_{\mathrm{RED}} + B_{\mathrm{MIR}})} \tag{4.8}$$

where B_{NIR} = brightness value of near-infrared band, B_{RED} = brightness value of RED band, B_{GREEN} = brightness value of green band, B_{MIR} = brightness value of middle-infrared band.

These approaches could also be executed directly from Google Earth Engine (GEE). GEE allows to directly compute any such indices for individual year as well as a range of years like a season, pre-dam and post-dam periods. This is very time effective approach and incorporates all the available images in a define period of time within an archive. Even multisensory sources could also be used together. Here wetland of all along time period defined could be merged and presented accurately. As a result, areal extent shows relatively higher than any single year. In the present case, wetland map could be developed based on pre- and post-dam periods. However, areal extent of wetland could not be significantly different. This map could be used for assessing wetland health based on multi-dimensional aspects.

4.2.1.1 Accuracy Assessment of Wetland Mapping

Overall accuracy, the Kappa coefficient, and the Receiver Operating Characteristics (RoC) curve are among the most widely used methods for cross-validating wetland maps (Zhu and Gong 2014; Zheng et al. 2015). The reliability of such assessments strongly depends on the quality of ground control points, where factors such as their density, spatial distribution, and spacing play a crucial role in ensuring robustness. In the present case, a total of 143 ground control sites were utilized for accuracy evaluation.

$$\text{Overall accuracy} = \frac{\text{Total number of correct samples}}{\text{Total number of samples}} \times 100 \tag{4.9}$$

$$\text{Kappa Statistics } (K) = \frac{(P_T \times P_{TC}) - \sum(T_C \times T_R)}{P_T^2 - \sum(T_C \times T_R)} \tag{4.10}$$

where P_T = total reference pixel, P_{TC} = total corrected classified pixel, T_C = total pixel for each class, and T_C = total reference pixel for each class.

4.2.2 Classifier-Based Approach

Multi-spectral (MS) satellite images from different sources are the basis of image classification. Two basic approaches are used for image classification, i.e., (1) unsupervised image classification and (2) supervised image classification. The second approaches are more rigorously used for this purpose. In relevant image processing software, spectral/spatial signature-based training to computer could help to support classification. For example, suitable signatures for water body from MS image with adequate number could help to classify segregate wetland from other forms of land use/land cover. Different classifiers are available for classification like maximum likelihood classifier, parallelepiped classifier, feature space classifier, etc. Suitable one could be adopted. In the present case, maximum likelihood classifier-based approach was applied for land cover classification. Here along with wetland, other land cover was also mapped but if not needed others could be escaped.

4.2.3 Machine Learning (ML) and Deep Learning (DL) Approaches

Machine learning (ML) and deep learning (DL) algorithms are drawing attention to the scholars for different purposes, including wetland demarcation. A good many algorithms like artificial neural network (ANN), convolutional neural network (CNN), Naive Bayes (NB), random forest (RF), k-nearest neighbors (KNNs), classification and regression tree (CART), and many more are available for this purpose. Basic mechanism behind this is to train algorithms using suitable training sites. Hence, the success of this work depends on accuracy of training sites and precision of training the algorithm. Accurate collection of training sites from field or any other relevant secondary sources is essential for this. Malinowski et al. (2020), Jamali (2021) revealed that ML and DL approaches could accurately classify wetland from a complex array of land use/land cover. In this present work, an example of the application of such approaches for wetland delimitation was shown. Very recently Google Earth Engine (GEE) embedded ML was also used for the same purpose.

4.2.4 Definitional Criteria-Based Approach

Keddy (2010) characterized wetland in reference to three major criteria, including (1) presence of water, (2) hydrophytes, and (3) hydric soil. Water presence may be perennial, semi-perennial, and ephemeral. The depth of water may be deeper water to shallow to saturated soil. *Hydrophytes* are the vascular and non-vascular plants that live in aquatic condition particularly in submerged or partially emerged or saturated conditions. *Hydric soil* is the saturated soil, seasonally or perennially submerged (anaerobic condition) soil supports hydrophytes. If a water body satisfies these three conditions is called wetland. So field investigation regarding these parameters could help to delineate wetland.

4.3 Method for Simulating Wetland Area

Forecasting future wetland extent serves as an essential tool in shaping conservation strategies and anticipating changes in habitat structure and ecological conditions (Ji et al. 2025). In regions like the present study area—where significant wetland degradation has already occurred (Pal and Saha 2018; Pal and Sarda 2020)—such predictive approaches become particularly vital for guiding sustainable landscape management. Numerous prior studies have applied integrated models such as the artificial neural network (ANN)–cellular automata (CA) framework to forecast spatial transformations in land cover (Rezaeianzadeh et al. 2018). ANN, recognized for its strength in modeling nonlinear and complex relationships between input variables and target outcomes (Kohail and El-Halees 2011), has outperformed several other models in terms of prediction accuracy (Rahman and Esha 2022). In the ANN-CA approach, the simulation incorporates environmental control variables—key deterministic factors that influence land cover transitions—which lead to improved alignment between actual and predicted outcomes (Aguilera et al. 2016; Zamani and Khairi 2018). Recently, ANN models have also gained popularity in the domain of water resources modeling and ecosystem analysis (Chen et al. 2020; Latif et al. 2021). Given the utility of this modeling approach in integrating crucial land transformation drivers, the current study included variables such as vegetation expansion, agricultural encroachment (Saha and Pal 2019), and increasing built-up areas (Cai et al. 2012). These factors are particularly relevant within the study's floodplain context, where such land use pressures significantly affect wetland dynamics (Saha and Pal 2019). To represent these conditioning variables, satellite-derived indices were prepared, including the Normalized Difference Vegetation Index (NDVI) to delineate vegetated and agricultural land, and the Normalized Difference Built-up Index (NDBI) to map urbanized areas. These index layers were generated for base years and projected for future scenarios. Model inputs were validated using reference datasets obtained from both field surveys and high-resolution satellite imagery available via Google Earth, utilizing 130 validation points for NDVI and 270 for NDBI. Validation was performed using the Kappa coefficient, ensuring reliability of the classification.

4.3.1 Accuracy Assessment of the Simulation Models

A critical component of any predictive modeling exercise is the evaluation of its accuracy. As emphasized by El Naqa et al. (2018), without a rigorous assessment of a model's performance, its scientific relevance remains limited. In the current study, direct validation of future wetland extent is inherently not feasible. Therefore, to gauge the reliability of predictions, the model was first used to simulate the wetland extent for the year 2020, and this was compared to the actual wetland condition observed in the same year. If this comparison yielded acceptable results, predictions for 2025 and 2030 were deemed credible. To assess the comparative performance and sensitivity of different simulation models, Receiver Operating Characteristic (ROC) curves were used, and the Area Under the Curve (AUC) was calculated. The ROC analysis helps determine a model's classification performance by plotting the true positive rate (sensitivity) against the false positive rate (1—specificity), as outlined by Fawcett (2006). This technique is widely adopted for evaluating the strength of model predictions in spatial and ecological modeling (Pourghasemi et al. 2013). Interpretation of AUC values is standardized: scores between 0.90 and 1.00 indicate excellent predictive power, 0.80 to 0.90 denote good, 0.70 to 0.80 are considered fair, and values between 0.60 and 0.70 fall under poor accuracy levels (Rasyid et al. 2016). An AUC of 1.0 represents a perfect classification model, 0.5 indicates no better than random chance, and 0 implies complete misclassification (Hiestermann and Rivers-Moore 2015).

4.4 Result

4.4.1 Wetland Map Based on Spectral Water Indices

Figure 4.1 illustrates the seasonal wetland distribution across selected years, derived using the NDWI for both pre-monsoon and post-monsoon periods. The figure highlights the temporal and seasonal variability in wetland extent. During the pre-dam era, the reduction in wetland area was relatively moderate; however, the rate of wetland contraction increased notably following the construction of the dam, indicating a stronger post-dam impact on seasonal wetland dynamics.

For instance, in 2000, the post-monsoon wetland area was 1680 km^2 which was reduced to 802 km^2 in 2013. So within this period more than 50% wetland was lost (Fig. 4.1) and it is very thought-provoking. In 2015 and 2024, wetland areas were respectively 413 km^2 and 124 km^2 showing continuation of wetland squeeze. In pre-monsoon also wetland area loss is very prominent. Seasonal change of wetland area is very significant. For example, in 1991, about 256 km^2 area was under wetland domain in pre-monsoon season and the area was expanded to 744 km^2 in post-monsoon showing about three time expansion. In post-dam time, this seasonal fluctuation is considerably greater than pre-dam (five times). Here it is noteworthy to mention that since the wetland in this area is majorly fed by rainfall and river spill water. About 80% rainfalls happen during monsoon months (June to September). So wetland extent is expected to be high in monsoon season. But due to lack of cloud-free image, immediately after monsoon season, images were collected

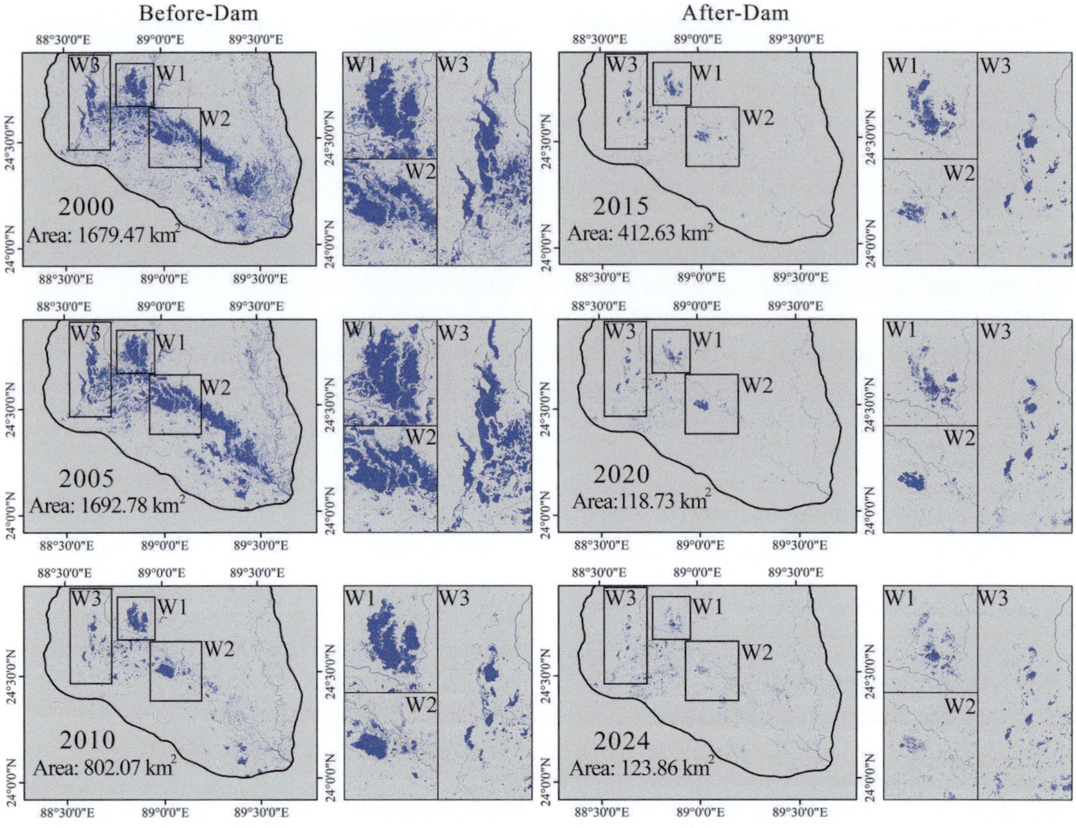

Fig. 4.1 Seasonal wetland maps for pre- and post-dam years using Normalized Difference Water Index (NDWI). Areas indicated in blue colour represents surface water bodies

and mapped. So, seasonal wetland dynamics is not as same as done here. Here it is further mentionable that in figure only wetland scenario is presented for six years, but for the sake of other analysis, wetland maps were prepared for other years.

4.4.1.1 Validation of the Wetland Map

For validating wetland maps (NDWI images), overall accuracy and Kappa coefficient were computed and displayed in Table 4.2. The outcomes show that in all the years' overall accuracy was found > 89% and K value > 0.81 which indicate very good to excellent agreement between wetland map and ground reality (Monserud and Leemans 1992). Considering this, the wetland maps could be treated as valid.

4.4.2 Wetland Maps Based on Classifier Approach

Segregation of wetland from other land covers using supervised image classification is traditionally used. The present examples, entire case unit was not taken into account but it was applied over a part of the study unit in order to present the result which could be received from this approach. Figure 4.2 presents the maximum likelihood classifier-based classification of land use/land cover. From this it is evident that wetland part is well detected. Google Earth image and ground truth (34 sites)-based cross-validation using Kappa coefficient ($K = 0.79$ in 1998 and 0.81) confirmed its validity. From the two maps, reclamation of wider wetland area into seasonal agricultural land was reported.

Table 4.2 Overall accuracy and Kappa co-efficient of wetland indicating NDWI water indices in pre- and post-monsoon seasons

Phase	Year	Pre-monsoon				Post-monsoon		
		Overall accuracy	Kappa co-efficient	Degree of agreement	Year	Overall accuracy	Kappa co-efficient	Degree of agreement
Pre-dam	1991	89	0.81	Very good	1990	94.25	0.91	Excellent
	1995	94	0.88	Excellent	1995	94	0.88	Excellent
	1999	91.33	0.82	Very good	2000	92	0.82	Very good
	2005	93.77	0.87	Excellent	2005	94.67	0.89	Excellent
	2010	95.67	0.91	Excellent	2010	96.33	0.92	Excellent
Post-dam	2014	92.6	0.85	Excellent	2013	93.67	0.87	Excellent
	2015	91	0.81	Very good	2015	95	0.9	Excellent
	2017	92.33	0.85	Excellent	2017	93	0.87	Excellent
	2019	96.46	0.95	Excellent	2020	98.4	0.96	Excellent
	2021	94.44	0.92	Excellent	2021	95.6	0.91	Excellent
	2024	93	0.85	Excellent	2024	94	0.87	Excellent

Note If kappa co-efficient value is > 0.85 = excellent agreement; 0.7–0.85 = very good agreement; 0.55–0.70 = good agreement; 0.40–0.55 = fair agreement; < 0.4 = poor agreement (Monserud and Leemans 1992)

4.4.3 Wetland Map Based on ML Approach

Of late machine learning (ML)-based image classification is very popular and often used for wetland classification. Random forest (RF), artificial neural network (ANN), support vector machine (SVM), classification and regression trees (CARTs), etc., are some of the common supervised classification techniques. Here as examples, minimum distance, SVM, RF, and CART were applied for wetland mapping from Landsat image, 2024. CART is a tree-based framework that has been widely used in remote sensing applications. Efficient training could produce robust classification. Total 200 training sites were used for wetland classification. Figure 4.3 shows the classified wetland maps as of post-monsoon, 2024 using all the ML classifiers. All the maps showed significant variation of wetland extent but classification accuracy is acceptable in most of the cases on the basis. Receiver Operating Characteristics (RoC) curve (area under curve or AUC value ranges from 0.90 to 0.95) which were computed showing the classification accuracy. Minimum distance, SVM, RF, and CART-based image classification respectively predicted 621.52 km^2, 1683.68 km^2, 1932.02 km^2, and 2135.31 km^2 areas as wetland. There is similarity in wetland areas among the applied models except minimum distance. But this does not imply the minimum distance classifier would not be outperforming. Google-based 139 testing sites were used for validation of wetland. Kappa coefficient (K) was computed for this (Eq. 4.10). Computed K value ranges from 0.61 and 0.8 with the highest accuracy in the case of minimum distance classifiers and hence could be treated outperforming.

4.4.4 Trends in Wetland Area

Table 4.3 presents the trend of wetland area from 1990 to 2011 and from 2012 to 2024, representing the pre- and post-dam phases, respectively. In the pre-dam phase, no definite trend was detected in either of the two seasons considered. For the pre-monsoon season, the computed least squares equation is $y = -1.8203x + 3806.2$, with an r^2 value of 0.02. In the post-monsoon season, no significant trend was observed. However, after damming, the situation changed. In both seasons, statistically significant declining trends

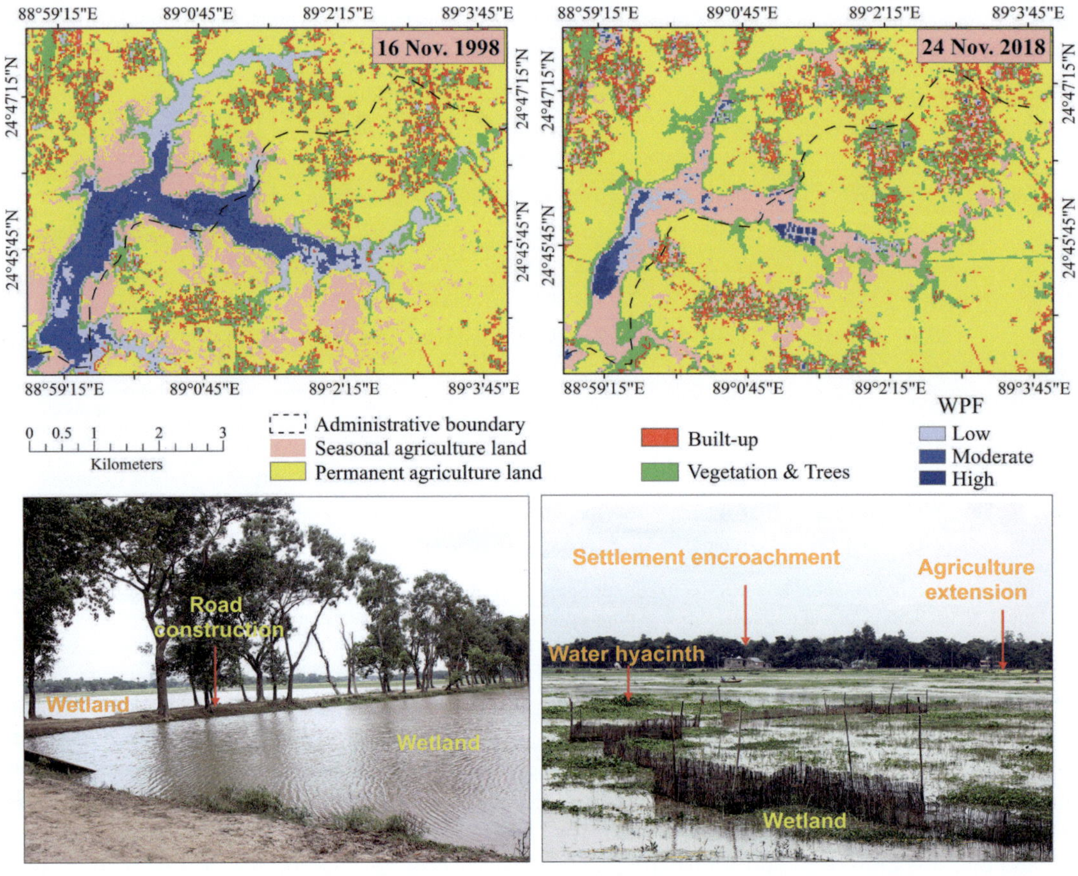

Fig. 4.2 Maximum likelihood classifier-based land use/land cover maps of Raktadaha Wetland (Beel) in lower reach of Atreyee River basin

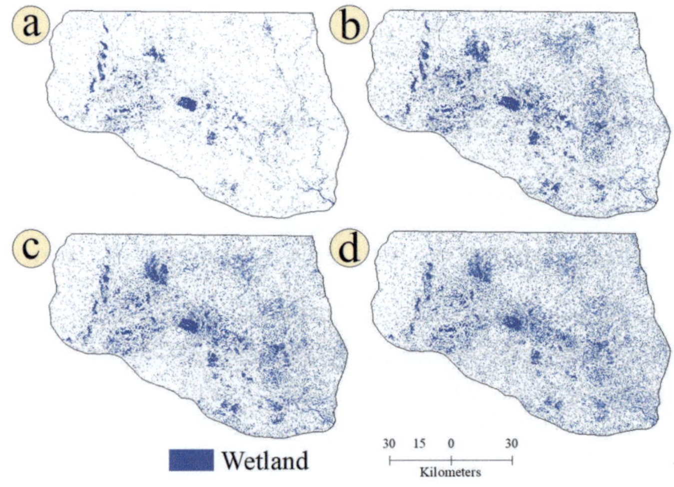

Fig. 4.3 Wetland area predicted based on machine learning-based classifiers including **a** minimum distance, **b** SVM, **c** RF, and **d** CART

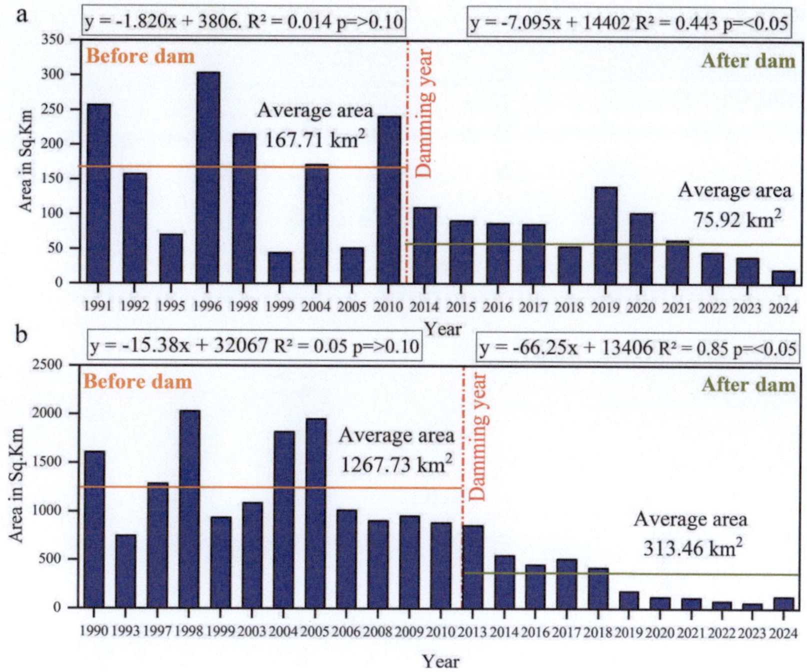

Fig. 4.4 Trends in wetland area of the Atreyee River basin **a** pre-monsoon season and **b** post-monsoon season

Table 4.3 Seasonal trends of wetland area in the Atreyee River basin, 1990–2024 (values marked with * are statistically significant at the 0.05 level)

Period	Year	Trend equation	r^2	CV (%)	IX
Pre-monsoon	1990–2011 (before dam)	$y = -1.8203x + 3806.2$	0.02	53.44	53.43
	2012–2024 (after dam)	$y = -7.0957x + 14{,}402$	0.44*	44.37	39.76
Post-monsoon	1990–2011 (before dam)	$y = -15.385x + 32{,}067$	0.05	34.67	34.63
	2012–2024 (after dam)	$y = -66.25x + 134{,}061$	0.85*	78.73	41.57

were detected. For example, the post-monsoon trend equation is $y = -66.25x + 134{,}061$, with a coefficient of determination (r^2) of 0.85 ($p < 0.05$). Year-wise wetland area plots and the trend values shown in Fig. 4.4 and Table 4.3 confirm that sharp areal shrinkage occurred in the post-dam period. When comparing average wetland areal changes between pre- and post-dam phases, in the post-monsoon season the wetland area reduced by about 955 km^2 (from 1268 to 313 km^2). In the pre-monsoon season, a reduction in areal variability was also detected. Graf (2006) reported similar findings in the case of dam-induced changes in river flow, which is equally true for wetland areal extent. Since river flow and wetland areal dynamics are strongly associated, such shrinkage is understandable. As highlighted in Chap. 2, a reduction in river flow variability was identified during the post-dam period. Under these conditions, with sparse water supply to wetlands, the chances of high-level fluctuations in wetland area remain very limited.

4.4.5 Simulated and Predicted Wetland Area

Figure 4.5 shows the geographical distribution of wetlands derived from NDWI indices and simulated and predicted wetland areas. Actual and simulated wetland areas were respectively 637.27 km^2 and 585.72 km^2 in 2010 and 428.63 km^2 and 498.13 km^2 in 2020. In both, the cases over prediction are observed. It is in fact due to the acceler-

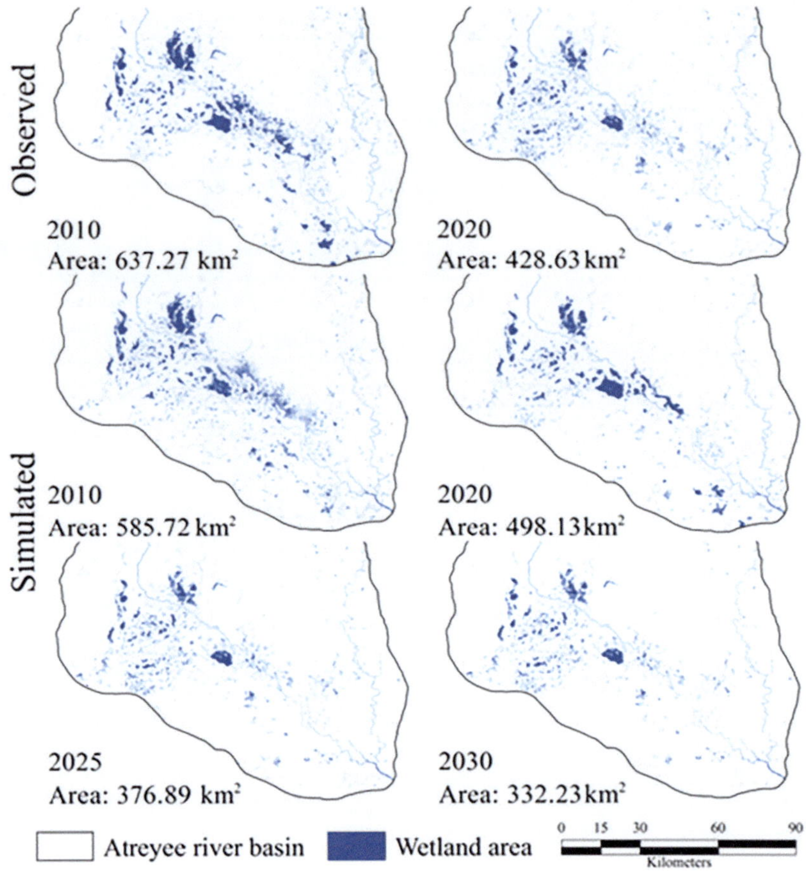

Fig. 4.5 Actual and predicted wetland areas (prediction has been done using CA-ANN)

Table 4.4 Actual and predicted areas in different years

Year	2010		2020		2025	2030
	Actual	Simulated	Actual	Simulated	Forecasted	
Area in km^2	637.27	585.72	428.63	498.13	377.87	332.23

ation of wetland loss over the progress of time. Based on historical data, when the future has been predicted, the historical rate of wetland change was taken as an input. Acceleration was not taken but increasing human pressure caused for growing loss of wetland. In 2025 and 2030, predicted areas are likely to be decreased to 377.87 km^2 and 332.23 km^2 (Table 4.4). If the spatial pattern of wetland loss is observed, it is very clear that small and fringe wetlands have lost their identity more prominently.

4.4.6 Validation of Simulated Wetland Area

For validating the simulated models of 2010 and 2020, these are compared with the actual wetland area or areal extent of the respective years. Overall accuracy for the simulated models of 2010 and 2020 is greater than 90. Whereas, the Kappa coefficient and AUC under ROC are greater than 0.80 for simulated models (Table 4.5). From all these computed values, it can be treated that the simulated models are valid. If the simulated models are

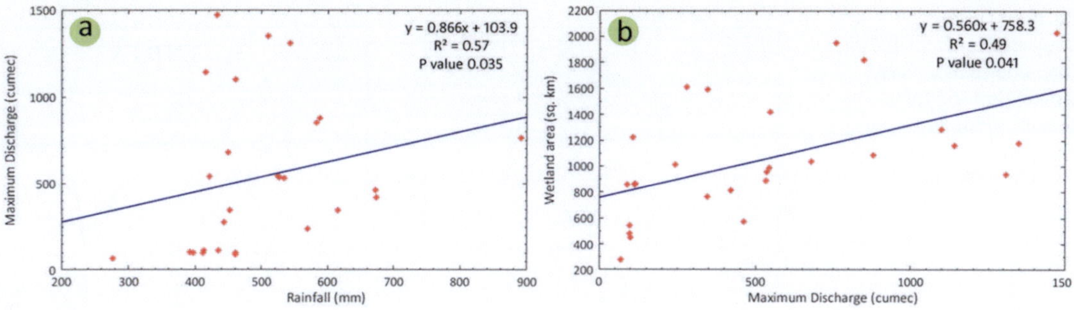

Fig. 4.6 Linear regression between **a** monthly average rainfall and maximum discharge **b** maximum discharge and wetland area

Table 4.5 Sensitivity analysis of simulated wetland areas for the years 2010 and 2020

Model	2010			2020		
	Overall accuracy	Kappa coefficient	ROC-AUC	Overall accuracy	Kappa coefficient	ROC-AUC
ANN-CA model	91.33	0.82	0.84	94	0.88	0.87

valid for 2010 and 2020, expectedly it is likely to be valid in the case of 2025 and 2030.

The result indicates that the future prediction of wetland areas by 2030 is squeezing. The main cause for growing this situation is inadequate water supply to the wetland from the river. Therefore water availability within wetlands gradually decreases. This episode exerts an impact on reducing wetland areas in the current study. The same result was found by Pal and Saha (2018) in their study over the Punarbhaba River adjacent to this case basin. Shallow and erratic wetlands have transformed into agricultural land at a very rapid pace. Commonly found shallow and seasonal wetland in this area provides a good opportunity to cultivate mainly paddy cultivation during the monsoon season. Many researchers documented the conflict between wetland areas and agricultural land in developing countries (Saha and Pal 2019; Islam 2016). Many of the tie channels have lost their water flow due to the lowering of water level in the main channel is another reason behind the squeezing of wetland areas in the present study (Pal and Saha 2018). This type of work could be treated as an effective tool for wetland management and restoration.

4.5 Discussion

The present chapter used multiple approaches for wetland inventory and finally adopted spectral water indices for further work. Overall, the present case reported reduction of wetland area both during pre- and post-dam periods. But significantly declining trend was detected in post-dam period. Since the floodplain wetlands are strongly based on river inundation, reduction in inundation and magnitude of inundation is a very vital reason behind wetland loss (Bassi et al. 2014). In subhumid region, river discharge is strongly controlled by rainfall. Figure 4.6 shows that rainfall change is not significant in the present study unit. Discharge reduction in spite of no such remarkable rainfall reduction implies that not rainfall, other intervening cause is crucial here. Dam-induced water diversion is a vital one along with other forms of water harvesting from river. To assess the interrelationship between annual wetland area and river discharge, a linear regression analysis was performed specifically for the post-monsoon season. The results indicated a positive correlation, suggesting that higher discharge volumes support the expansion of

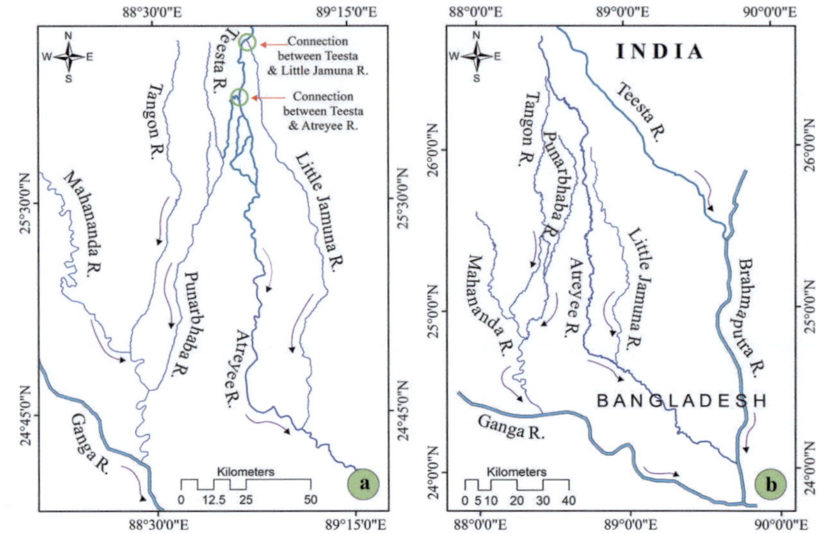

Fig. 4.7 Distract of Atreyee River from Teesta River course **a** drainage network in 1767 (after Rennell 1788) and **b** drainage network in 2017 (Google earth image)

wetland areas. The derived regression model, with an equation of $0.560x + 758.3$ and a coefficient of determination (R^2) of 0.49, confirms this relationship. The R^2 value is statistically significant at the $p < 0.05$ level, supporting the reliability of the model. Further analysis of the post-dam period revealed a substantial decline in monsoon-related hydrological characteristics. On average, monsoon discharge in the Atreyee River was reduced by approximately 65.7%, while the frequency of floods exceeding danger level dropped by 90%. Additionally, peak monsoon discharges declined by 75.91%, reflecting a dramatic shift in river flow dynamics following regulation.

This explanation of dam-induced discharge reduction and wetland area squeeze is very true for post-dam condition. But why wetland was also reduced during pre-dam condition?

In order to understand the fact, it would be better to trace back the history of drainage evolution of this area. Teesta River of North Bengal coming from Eastern Himalayan part was flowed through three rivers, including Atreyee, Punarbhaba, and Yamuna. Before 1787, these flows were continued. In 1787, intense rainfall led to a sudden and massive deposition of sand that obstructed the Atrai (Atreyee) River channel. This was followed by significant neo-tectonic movements during the eighteenth century, which redirected the course of the Teesta River toward the southeast (Debnath 2007). As a result of these events, the Atreyee River lost its connection to the glacially-fed Teesta system, cutting off its primary source of water (Das and Pal 2017). This incident turned the Atreyee River in to be-headed river (Fig. 4.7) converted it from perennial to semi-perennial. Flow volume in Atreyee was naturally reduced. This historical lineage is caused for dwindling spilling frequency and magnitude as well as wetland inundation.

In many parts of the wetland, reduced river discharge has led to an increasing reliance on rainfall as the primary water source. As a result, surface water tends to accumulate only in the deeper sections of wetland depressions, while the remaining areas often remain dry, except during periods of intense rainfall. The higher, peripheral zones of these wetlands typically remain either partially saturated for short periods or entirely dry throughout the year. Over time, this has transformed large portions of these wetlands into low-frequency water retention zones. This issue is particularly severe during the non-monsoon months, when dam operations divert river water for irrigation. Along the river, numerous government and privately managed water extraction schemes further exacerbate the problem. In the Indian section alone, between Samjia and Chak Bhatshal, there are 55 major River Lift Irrigation

(RLI) projects currently operational. The cumulative effect of water diversion and regulation has led to a noticeable decline in wetland extent and severely disrupted wetland-dependent ecosystems, especially those associated with the floodplain. Further contributing to this degradation are embankments, which restrict the natural overflow of the river into adjacent low-lying areas. On the Indian side of the Atreyee River, a 16-km-long embankment has been constructed, artificially elevating the natural flood spill level by 2 to 3 m (Pal 2016). Additionally, tie channels—once vital for hydrological connectivity between the river and wetlands—have been progressively encroached upon for agriculture and other land uses. These channels now carry water only during exceptional spill events. As they remain dry for consecutive years, local communities often repurpose them for cultivation, a pattern also observed in the Hizole wetland and surrounding areas of Murshidabad (Pal and Akoma 2009; Pal 2016). Another critical factor is the rapid decline of groundwater levels in the floodplain, primarily due to extensive extraction from underground aquifers. This has resulted in a disconnection between surface water and the subsurface water table, further weakening the wetland's ability to retain water and sustain its ecological functions.

References

Aguilera H, Moreno L, Wesseling JG, Jiménez-Hernández ME, Castaño S (2016) Soil moisture prediction to support management in semiarid wetlands during drying episodes. CATENA 147:709–724

Ahmad M, Ahmad WS, Ahmad SN, Jamal S, Saqib M (2024) Tracing the roots of wetland degradation in India: a systematic review of anthropogenic drivers, ecological consequences and conservation strategies. GeoJournal 89(1):24

Bassi N, Kumar MD, Sharma A, Pardha-Saradhi P (2014) Status of wetlands in India: a review of extent, ecosystem benefits, threats and management strategies. J Hydrol Reg Stud 2:1–19

Cai Y, Zhang H, Pan W, Chen Y, Wang X (2012) Urban expansion and its influencing factors in natural wetland distribution area in Fuzhou City, China. Chin Geogra Sci 22:568–577

Chen C, He W, Zhou H, Xue Y, Zhu M (2020) A comparative study among machine learning and numerical models for simulating groundwater dynamics in the Heihe River Basin, northwestern China. Sci Rep 10(1):3904

Das RT, Pal S (2017) Exploring geospatial changes of wetland in different hydrological paradigms using water presence frequency approach in Barind Tract of West Bengal. Spat Inf Res 25:467–479

Debanshi S, Pal S (2020) Wetland delineation simulation and prediction in deltaic landscape. Ecol Ind 108:105757

Debnath A (2007) Changing course of Kameng River in the lower reaches (Doctoral dissertation)

El Naqa I, Ruan D, Valdes G, Dekker A, McNutt T, Ge Y et al (2018) Machine learning and modeling: data, validation, communication challenges. Med Phys 45(10):e834–e840

Fawcett T (2006) An introduction to ROC analysis. Pattern Recogn Lett 27(8):861–874

Feyisa GL, Meilby H, Fensholt R, Proud SR (2014) Automated water extraction index: a new technique for surface water mapping using Landsat imagery. Remote Sens Environ 140:23–35

Graf WL (2006) Downstream hydrologic and geomorphic effects of large dams on American rivers. Geomorphology 79(3–4):336–360

Gupta AK, Goyal MK, Singh SP (eds) (2023) Ecosystem restoration: towards sustainability and resilient development. Springer Nature

Hiestermann J, Rivers-Moore N (2015) Predictive modelling of wetland occurrence in KwaZulu-Natal, South Africa. S Afr J Sci 111(7–8):1–10

Islam SN (2016) Deltaic floodplains development and wetland ecosystems management in the Ganges–Brahmaputra–Meghna Rivers Delta in Bangladesh. Sustain Water Resour Manag 2:237–256

Jamali A (2021) Improving land use land cover mapping of a neural network with three optimizers of multi-verse optimizer, genetic algorithm, and derivative-free function. Egypt J Remote Sens Space Sci 24(3):373–390

Ji P, Su R, Wu G, Xue L, Zhang Z, Fang H et al (2025) Projecting future wetland dynamics under climate change and land use pressure: a machine learning approach using remote sensing and Markov chain modeling. Remote Sensing 17(6):1089

Jones HG, Vaughan RA (2010) Remote sensing of vegetation: principles, techniques, and applications. OUP Oxford

Keddy PA (2010) Wetland ecology: principles and conservation. Cambridge University Press

Kohail SN, El-Halees AM (2011) Implementation of data mining techniques for meteorological data analysis. Int J Inf Commun Technol Res 1(3)

Lacaux JP, Tourre YM, Vignolles C, Ndione JA, Lafaye M (2007) Classification of ponds from high-spatial resolution remote sensing: application to Rift Valley Fever epidemics in Senegal. Remote Sens Environ 106(1):66–74

Latif SD, Ahmed AN, Sathiamurthy E, Huang YF, El-Shafie A (2021) Evaluation of deep learning algorithm for inflow forecasting: a case study of Durian

Tunggal reservoir, Peninsular Malaysia. Nat Hazards 109(1):351–369
Ma S, Zhou Y, Gowda PH, Dong J, Zhang G, Kakani VG et al (2019) Application of the water-related spectral reflectance indices: a review. Ecol Ind 98:68–79
Malinowski R, Lewiński S, Rybicki M, Gromny E, Jenerowicz M, Krupiński M et al (2020) Automated production of a land cover/use map of Europe based on Sentinel-2 imagery. Remote Sens 12(21):3523
McFeeters SK (1996) The use of the Normalized Difference Water Index (NDWI) in the delineation of open water features. Int J Remote Sens 17(7):1425–1432
Monserud RA, Leemans R (1992) Comparing global vegetation maps with the Kappa statistic. Ecol Model 62(4):275–293
Pal S (2016) Impact of water diversion on hydrological regime of the Atreyee river of Indo-Bangladesh. Int J River Basin Manag 14(4):459–475
Pal S, Akoma OC (2009) Water scarcity in wetland area within Kandi Block of West Bengal: a hydro-ecological assessment. Ethiop J Environ Stud Manag 2(3)
Pal S, Saha TK (2018) Identifying dam-induced wetland changes using an inundation frequency approach: the case of the Atreyee River basin of Indo-Bangladesh. Ecohydrol Hydrobiol 18(1):66–81
Pal S, Sarda R (2020) Damming effects on the degree of hydrological alteration and stability of wetland in lower Atreyee River basin. Ecol Ind 116:106542
Pal S, Sarkar R, Saha TK (2022) Exploring the forms of wetland modifications and investigating the causes in lower Atreyee river floodplain area. Eco Inform 67:101494
Pourghasemi HR, Moradi HR, Fatemi Aghda SM (2013) Landslide susceptibility mapping by binary logistic regression, analytical hierarchy process, and statistical index models and assessment of their performances. Nat Hazards 69:749–779
Rahman MTU, Esha EJ (2022) Prediction of land cover change based on CA-ANN model to assess its local impacts on Bagerhat, southwestern coastal Bangladesh. Geocarto Int 37(9):2604–2626
Rasyid AR, Bhandary NP, Yatabe R (2016) Performance of frequency ratio and logistic regression model in creating GIS based landslides susceptibility map at Lompobattang Mountain, Indonesia. Geoenvironmental Disasters 3(1):19
Rezaeianzadeh M, Kalin L, Hantush MM (2018) An integrated approach for modeling wetland water level: application to a headwater wetland in coastal Alabama, USA. Water 10(7):879
Saha TK, Pal S (2019) Emerging conflict between agriculture extension and physical existence of wetland in post-dam period in Atreyee River basin of Indo-Bangladesh. Environ Dev Sustain 21:1485–1505
Saha TK, Pal S, Sarkar R (2021) Prediction of wetland area and depth using linear regression model and artificial neural network based cellular automata. Eco Inform 62:101272
Shen L, Li C (2010) Water body extraction from Landsat ETM+ imagery using adaboost algorithm. In: 2010 18th International conference on geoinformatics, pp 1–4. IEEE
Singh M, Sinha R (2022) A basin-scale inventory and hydrodynamics of floodplain wetlands based on time-series of remote sensing data. Remote Sens Lett 13(1):1–13
Singh M, Sinha R, Mishra A, Babu S (2022) Wetlandscape (dis) connectivity and fragmentation in a large wetland (Haiderpur) in west Ganga plains, India. Earth Surf Process Landf 47(7):1872–1887
Sinha R, Tandon SK (2014) Indus-Ganga-Brahmaputra plains: the alluvial landscape. In: Landscapes and landforms of India, pp 53–63
Wilson EH, Sader SA (2002) Detection of forest harvest type using multiple dates of Landsat TM imagery. Remote Sens Environ 80(3):385–396
Xu H (2006) Modification of normalised difference water index (NDWI) to enhance open water features in remotely sensed imagery. Int J Remote Sens 27(14):3025–3033
Zamani NW, Khairi SSM (2018) A comparative study on data mining techniques for rainfall prediction in Subang. In: AIP conference proceedings, vol 2013, no 1. AIP Publishing
Zheng Y, Niu Z, Gong P, Wang J (2015) A database of global wetland validation samples for wetland mapping. Sci Bull 60:428–434
Zhu P, Gong P (2014) Suitability mapping of global wetland areas and validation with remotely sensed data. Sci China Earth Sci 57:2283–2292

5 Damming Effect on Wetland Hydrological Strength

Abstract

This chapter focused on modeling the hydrological strength of wetlands using various machine learning algorithms, including the reduced error pruning tree (REPTree), artificial neural network (ANN), and random forest (RF). The analysis incorporated key hydrological indicators such as water presence consistency, water depth, and hydro-period to evaluate wetland functionality. By comparing modeled hydrological conditions before and after dam construction, the study aimed to quantify the impact of regulation on wetland water dynamics. Results from parameter-based analysis indicated a substantial decline in wetland categories following damming. Specifically, areas classified as perennial wetlands decreased by approximately 22%, while semi-perennial wetlands declined by around 47%. A particularly notable reduction was observed in post-monsoon high water presence zones, which dropped sharply from 467 km^2 to just 40.34 km^2. Among the machine learning models employed, all demonstrated reliable performance when compared with observed field conditions. However, the REPTree model emerged as the most accurate. According to its outputs, the proportion of the area classified under high hydrological strength decreased from 46.61% in the pre-dam period to 31.76% after regulation. This reduction was especially pronounced along the fringe zones of wetland ecosystems, indicating a spatially uneven decline in hydrological resilience due to dam-induced flow alterations.

Keywords

Water presence consistency · Hydro-period · Water depth · Modeling hydrological strength · Machine learning · Change of hydrological strength

5.1 Hydrological Alteration and Its Impact on Wetland

Wetland hydrological alteration refers to changes in hydrological conditions as a whole or in reference to the hydrological components (Shaffer et al. 1999). Change in hydrological condition is very natural. But if this change occurs abnormally, it becomes a major concern. When a wetland undergoes hydrological alterations, it inevitably leads to changes in its ecological functions and the sustainability of its natural resources (Orimoloye et al. 2020). In this context, understanding wetland hydrology has become increasingly critical—not only for the conservation of wetland systems but also for the protection of biodiversity that depends on them. As the world grapples with the combined challenges of climate change, water scarcity, and the rapid decline in ecosystem biodiversity, the significance of safeguarding these natural systems

S. Pal et al., *Damming and Wetland Eco-hydrology*, GIScience and Geo-environmental Modelling,
https://doi.org/10.1007/978-3-032-11729-8_5

has grown substantially. These issues are now recognized as some of the most pressing global threats to human well-being (World Economic Forum 2019). Wetland being a natural hydrological component with utmost ecological importance can help to reduce and overcome such crises up to a certain extent (Ma et al. 2021). Wetlands contribute critically to a range of essential functions, including flood mitigation, groundwater replenishment, carbon storage, drinking water provision, microclimate regulation, and various other ecosystem and socioeconomic services (Vári et al. 2022). These functions play a vital role in enhancing human well-being and promoting sustainable development, particularly by reducing ecological vulnerabilities and economic uncertainties (King et al. 2021).

In developing countries, the floodplain wetlands have been facing extensive hydrological pressure (Pal and Paul 2021). The floodplain wetlands are often characterized by active hydro-habitat conditions supported by the neighboring perennial channels and seasonal inundation (Tsoi et al. 2022). These wetlands are oxbow lakes, left channels, swamps, marsh, bog, fen, and sloughs (Hong et al. 2021). Riparian areas of Ganga, Brahmaputra, Bhagirathi, Padma, and its tributaries of Indian and Bangladesh act as emerging and nourishing centers for numerous floodplain wetlands (Khan and Rahman 2022). Agriculture borne wetland change is very evident in floodplain areas. But hydrological change particularly its component is very noteworthy. Some of the important components are water depth, seasonal fluctuation, wetted area, consistency of water appearance, hydro-period, etc. Changes can also be studied in reference to such individual parameters or in integration of all. Combine study of hydrological components can provide overall situation of hydrological condition. This condition could be treated as hydrological strength or water richness. High strength or water rich means the study unit is characterized by higher consistency, longer hydro-period, shallower water depth, lower level of water fluctuation, etc.

While dealing with compositing the parameters, suitable method selection is very vital. If methodological spectrum in this regard is reviewed, there is a wide array of modeling techniques. Machine learning and deep learning methods for modeling have received significant attention due to its efficacy and robustness (Ahmed et al. 2023).

Considering the background of the issue and methods, this chapter tried to model hydrological strength (HS) based on the dominant hydrological components.

5.2 Materials

Post-monsoon season Landsat satellite imagery was obtained from the US Geological Survey's earth explorer platform. This study utilized multi-temporal Landsat data spanning the years 1990 to 2024, all with a spatial resolution of 30 m, to support the analysis. The details of the images are mentioned in the material section of Chap. 4.

5.3 Methods

5.3.1 Preparing Data Layers for Developing Wetland Hydrological State Models

5.3.1.1 Assessing Hydrological Consistency Using the WPF Method

Hydrological consistency in seasonal wetlands refers to the regularity with which a wetland experienced inundation during specific periods over multiple years. To evaluate this, the WPF approach proposed by Borro et al. (2014) was employed. This method quantified how frequently water appeared in a wetland, thereby offering insights into the reliability of seasonal flooding. The analysis was carried out for both the pre-dam period (1990–2011) and the post-dam period (2012–2024), considering both pre-monsoon and post-monsoon seasons. The procedure involved converting NDWI-derived water body maps into binary format, where water pixels were assigned a value of 1 and non-water pixels a value of 0. These binary images were then aggregated phase-wise, following the procedure outlined in Eq. 5.1. This technique had previously been applied successfully in studies of dynamic wetlands by Mondal and Pal (2018) and Sarda and Das (2018). WPF values were expressed as percentages, ranging from 0 to 100. A WPF value close to 100 indicated high consistency in seasonal inundation, whereas values near 0 signified strong inconsistency.

$$\mathrm{WPF}_k = \frac{\sum_{i=1}^{N_t} P_W}{N_t} \times 100, \tag{5.1}$$

where WPF_k = Water presence frequency of pixel k; P_W = Pth pixel possessing water in the NDWI image; N_t = total number of images taken into consideration.

5.3.1.2 Estimating Wetland Water Depth Using Relative NDWI Intensity

The depth of water within a wetland is a crucial indicator of its capacity to deliver ecological functions (Duan and Niu 2018). One effective way to estimate water depth remotely was through NDWI calibration, which involved linking NDWI pixel values to observed field-based depth measurements. This method had been validated in several earlier studies (Donchyts et al. 2016; Ma et al. 2019; Pal and Sarda 2020). For the calibration in this study, average NDWI maps were generated for different time phases, and their spatial pixel value variations were analyzed to estimate relative differences in water depth. The 2019 NDWI map, corresponding to the post-dam period, was selected for calibration. This map was cross-referenced with 134 ground-measured water depth points to establish a relationship between NDWI values and actual water depth across the wetland landscape.

5.3.1.3 Seasonal Inundation Duration (Hydro-period) Analysis

The hydro-period (HP), or inundation duration, reflects the number of months each wetland pixel remained water-covered in a given year. This metric has been widely used to evaluate wetland seasonality and phenological behavior (Bertassello et al. 2019; Pal and Singha 2022). To generate hydro-period maps, monthly NDWI images were processed by converting them into binary maps, where water pixels were marked as "1" and non-water areas as "0." These were then integrated to compute the HP value for each pixel. Resulting values ranged from 0 to 12, where higher values (close to 12) indicated perennial wetland conditions, and lower values suggested semi-perennial or seasonal wetlands. For comparative analysis, hydro-period maps were developed for both the pre-dam and post-dam periods and were classified into three categories to assess changes in seasonal water retention at the pixel level. In highly seasonal riparian landscapes such as the study area, wetlands were particularly vulnerable to alternating wet and dry phases (Talukdar and Pal 2020). Consequently, hydro-period served as a critical indicator for evaluating hydrological strength. Although rainfall variability could also influence wetland behavior, it was excluded here due to the lack of spatial heterogeneity in rainfall distribution across the wetland zone.

5.3.2 Modeling Wetland Hydrological Strength with Machine Learning

5.3.2.1 Reduced Error Pruning Tree (REPTree)

The REPTree algorithm, a hybrid of decision tree (DT) and reduced error pruning (REP), offers an efficient machine learning solution for classification tasks (Quinlan 1987). This technique reorganizes the structure of the decision tree using a training dataset and applies pruning methods to simplify the tree, thereby improving its accuracy and reducing overfitting (Mohamed et al. 2012; Pham et al. 2019). Pruning can occur through pre-pruning, which is faster but less accurate, or post-pruning, which produces higher precision (Chen et al. 2019). In this study, post-pruning was adopted to generate a sequence of simplified tree structures and select the most accurate one based on validation with test data.

5.3.2.2 Artificial Neural Network (ANN)

The artificial neural network (ANN) is a widely applied and versatile machine learning model (Mosavi et al. 2018). Structurally, it comprises input, hidden, and output layers, which allow it to learn complex patterns within datasets (Cardinaux et al. 2020; Nguyen et al. 2020). ANNs are particularly useful in modeling nonlinear relationships and are capable of adjusting internal parameters dynamically through the hidden layer (Choi et al. 2019). In this study, the multilayer perceptron (MLP) topology was used, trained using the back-propagation algorithm (Zhao et al. 2019). This feed-forward architecture enables the ANN to classify data based on a comprehensive training set (Paola and Schowengerdt 1995). Key inputs such as WPF, relative water depth, and hydroperiod were used to construct a reliable model for wetland hydrological stability assessment.

5.3.2.3 Random Forest (RF)

The random forest (RF) model, introduced by Breiman (2001), is a popular ensemble-based classifier that integrates multiple decision trees to achieve higher prediction accuracy (Youssef et al. 2016; Behnia and Blais-Stevens 2018). By employing bootstrap aggregation (bagging) and introducing randomness in feature selection, RF increases diversity among trees and reduces overfitting. The algorithm constructs multiple regression trees and aggregates their outputs to finalize predictions based on majority voting. In this study, RF was particularly effective due to the high-dimensional nature of the dataset, which allowed it to leverage feature combinations to accurately model the hydrological strength of wetland systems (see Table 5.1).

Table 5.1 Optimized parameters for machine learning models used in wetland hydrological strength assessment

Algorithm	Model settings and parameters
Artificial neural network (ANN)	• Number of hidden layers: 4 • Learning rate set to 0.25 • Momentum coefficient: 0.19 • Random seed value: 7 • Training iterations: 500 • Validation threshold: 20 • Applied binary filter conversion: enabled
Random forest (RF)	• Batch processing size: 150 • Seed value for reproducibility: 4 • Number of trees grown: 150 • Maximum depth of individual trees: 2 • Out-of-bag (OOB) error estimation: enabled • Attribute importance computation: enabled
Reduced error pruning tree (REPTree)	• Batch size: 150 • Seed: 3 • Maximum tree depth: 2 • Minimum instances per leaf: 2.0 • Minimum variance reduction threshold: 0.001 • Spread count initialization: enabled

5.3.3 Validation of Wetland Hydrological Strength Models

As noted by Nampak et al. (2014), the reliability of any predictive model is compromised if its accuracy and uncertainty are not properly evaluated. To validate the machine learning models used for wetland hydrological strength assessment, several performance metrics were applied. These included sensitivity (SE), specificity (SP), Kappa statistic (K), the area under the curve (AUC) of the receiver operating characteristic (ROC) curve, and the Matthews correlation coefficient (MCC). These indices offered robust statistical evaluation: values of SE, SP, K, and AUC typically ranged from 0 to 1, where values closer to 1 indicated higher predictive accuracy (Yang and Zhou 2015; Warrens and Pratiwi 2016). MCC was used to assess the performance of binary classifications, offering a comprehensive measure of model quality. Its values ranged from − 1, indicating complete disagreement, to + 1, reflecting perfect prediction agreement (Chicco and Jurman 2020).

5.4 Results

5.4.1 Hydrological Attributes for Developing Hydrological Strength of the Wetland

5.4.1.1 Water Presence Frequency

In seasonally inundated wetland, consistency of water presence is very crucial both hydrologically and ecologically. Figure 5.1 presents the

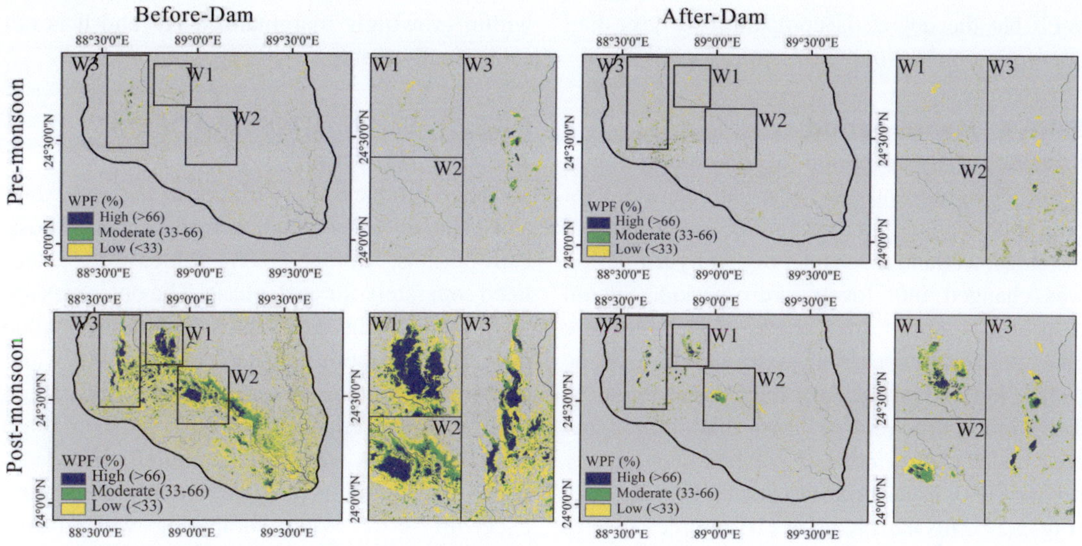

Fig. 5.1 Inundation frequency maps developed using water presence frequency (WPF) in pre-monsoon and post-monsoon seasons, in the **a** pre-dam **b** post-dam periods

spatio-seasonal water presence scenario of the pre and post-dam periods. WPF maps were categorized into three subclasses, i.e., high, moderate, and low indicating level of water presence. High WPF (> 67%) in a wetland pixel means in a selected study period, out of total years, at 67% or more times, that pixel secured water. In reference to spatial distribution of high or low frequencies, there is no definite trend of low to high or high to low from upper to lower reach of the study unit. High WPF areas are concentrated in some low lying wetland patches. Wider parts of lower reach were identified with very little concentration of high WPF wetland (Fig. 5.1). In reference to river, if lateral distribution is considered, high WPF areas are found at relatively proximate areas to main river. Consistency statistics between pre- and post-dam period clearly showed significant reduction of area under all the WPF levels.

Wetland area with high water presence consistency in post-monsoon season was significantly reduced from 467 to 40.34 km^2. About 2312 km^2 wetland area with lower WPF class was also considerably reduced to only 222 km^2 (Table 5.2). These statistics clearly presented degradation scenario of wetland. Conversion of wetland from high WPF category to lower classes shows growing inconsistency in WPF. Similarly, reduction of low WPF area signifies complete wipe out of wetland. In pre-monsoon season, there is also declining trend of consistency level of

Table 5.2 Seasonal areal extent of wetlands under different WPF classes in pre- and post-dam periods

WPF (%)	Pre-monsoon				Post-monsoon			
	Before dam		After dam		Before dam		After dam	
	Area in km^2	% of area	Area in km^2	% of area	Area in km^2	% of area	Area in km^2	% of area
Low	181.42	72.27	154.30	75.59	2312.11	69.29	221.62	64.63
Moderate	50.96	20.30	32.01	15.68	557.52	16.71	80.95	23.61
High	18.65	7.43	17.82	8.73	467.02	14.00	40.34	11.76
Total	251	100	204	100	3337	100	343	100

WPF but the degree is comparatively less than post-monsoon season.

5.4.1.2 Hydro-period

Scenario of hydro-period of the wetland was changed significantly in post-dam period. Not only total wetland area was reduced but most of the cases, wetland area under higher hydro-period was changed into lower hydro-period. Spatial distribution shows that outskirts wetlands are characterized by narrower hydro-duration. Some deeper wetland patches recorded higher duration. The windows mentioned clearly highlighted that area under clearly defined patches of relatively higher HP was turned into lower HP category (Fig. 5.2). This narrowing of HP enforces species to complete their life cycle, breeding system within growingly marginalized HP and it is not ecologically viable.

5.4.1.3 Seasonal Wetland Depth Dynamics

To analyze the seasonal variation in wetland extent and relative water depth across pre- and post-dam periods, average NDWI values were calculated separately for each phase. The outcomes are illustrated in Fig. 5.3, which highlights notable differences in both wetland coverage and water depth between the two seasons. A substantial portion of the wetland area is observed to undergo seasonal drying, with even perennially inundated zones experiencing a reduction in water depth exceeding 60%. When comparing wetland area losses between the two phases, it was found

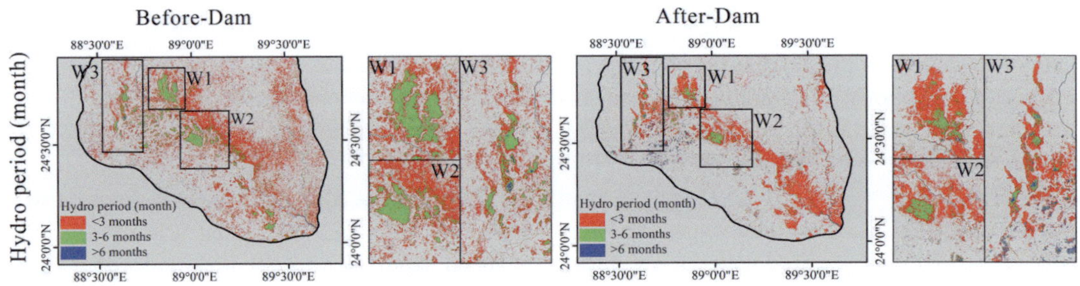

Fig. 5.2 Wetland maps showing the spatial pattern of hydro-period(HP) of HP in pre- and post-dam periods

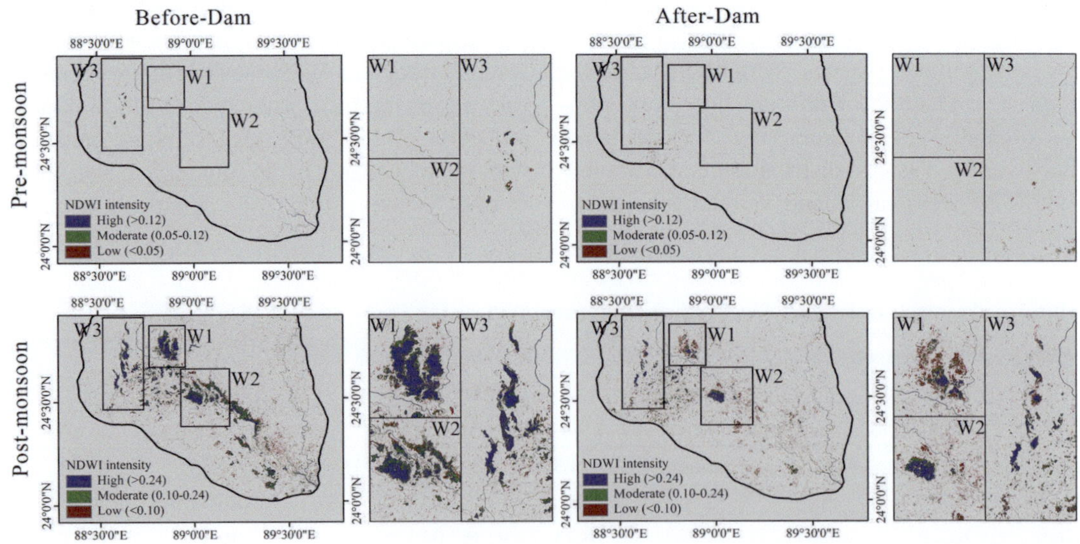

Fig. 5.3 Seasonal wetland condition using Normalized Difference Water Index (NDWI) intensity in pre and post-dam periods

Table 5.3 Season-wise areal extent of wetland area both during pre and post-dam periods

Season	NDWI sub-class	Before dam		After dam	
		Area in km^2	% of area	Area in km^2	% of area
Pre-monsoon	< 0.05 (Low)	19.92	43.56	13.61	64.50
	0.05–0.12 (Moderate)	16.75	36.63	7.27	34.46
	> 0.12 (High)	9.05	19.81	0.22	1.04
	Total	45.72	100	21.10	100
Post-monsoon	< 0.10 (Low)	418.65	48.04	41.22	45.28
	0.10–0.24(Moderate)	282.83	32.45	33.75	37.07
	> 0.24(High)	169.98	19.51	16.07	17.65
	Total	871.46	100	91.04	100

that perennial wetlands declined by 22%, and semi-perennial wetlands saw a reduction of 47%. Specifically, in the post-monsoon period, the area classified under low water depth conditions (NDWI < 0.1) dropped significantly—from approximately 419 km^2 to just 41 km^2, as shown in Table 5.3. This emphasizes that seasonal wetlands are particularly vulnerable to land use transformation and potential reclamation. Further analysis examined whether the intensity of seasonal fluctuations increased after dam construction. While the findings did not indicate any notable shifts in rainfall or temperature patterns during the post-dam period, widespread alteration in wetland characteristics was evident. Shallow, intermittent wetlands were disproportionately affected, with many being entirely lost. The data also revealed a transformation trend, where semi-perennial wetlands were frequently converted into non-wetland areas, and perennial wetlands were downgraded to either semi-perennial or non-wetland status following river regulation.

5.4.2 Modeling Hydrological Strength of the Wetland

Figure 5.4 illustrates the spatial distribution of wetland HS for both pre- and post-dam periods, as modeled using three machine learning algorithms: RF, ANN, and REPTree. During the pre-dam phase, areas with high hydrological strength were predominantly located near the main river channel. However, following dam construction, much of this high-strength zone experienced a substantial reduction. In the post-dam scenario, a significant share of the wetland area—previously characterized by strong water availability—was lost, particularly in downstream sections, where transformation was most severe. Large segments of wetlands situated along the river corridor also underwent noticeable changes, with many shifting from high to moderate or low HS categories. Table 5.4 summarizes the spatial extent of different HS zones identified by each machine learning model. Prior to damming, approximately 47% of the wetland area fell under the high HS category across all models. This figure dropped to about 32% in the post-dam period, with only minor differences observed between the models in terms of spatial coverage. Although the absolute area of the moderate HS zone decreased slightly (from 210 to 187 km^2), its relative proportion increased by 14%, largely due to the reclassification of formerly high-strength zones. Overall, the machine learning outputs showed strong alignment in delineating high, moderate, and low HS zones, both in terms of area and geographic placement. This suggests that all three models are reasonably consistent and reliable for hydrological strength analysis. Nevertheless, as with any classification model, further validation is recommended to determine the most effective and accurate approach.

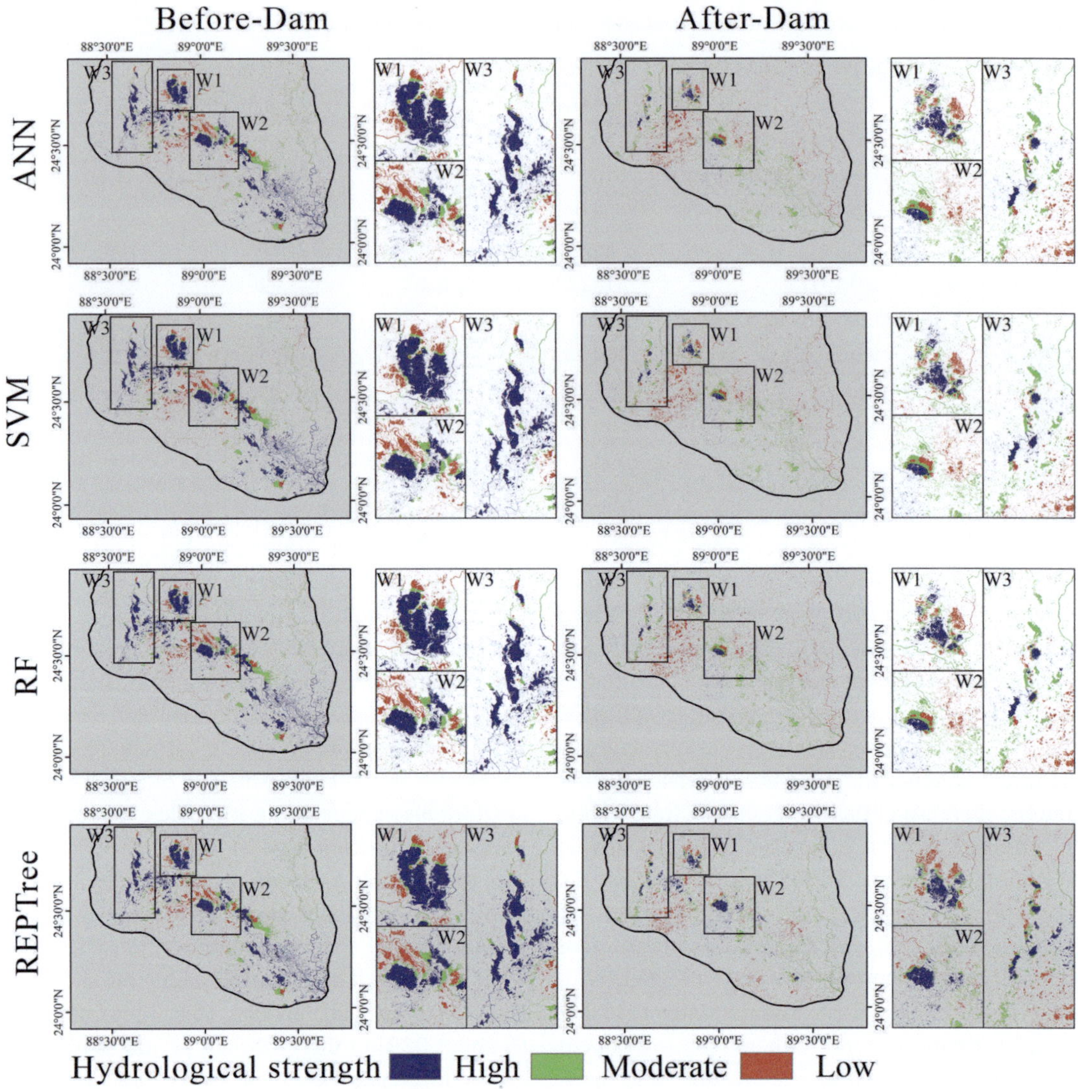

Fig. 5.4 Wetland hydrological strength models integrating hydrological parameters for pre-dam (1990–2011) and post-dam (2012–2024)

5.4.3 Accuracy Assessment of the Wetland Hydrological State Model

The results from various statistical accuracy metrics clearly indicate that all the machine learning models used in this study exhibited strong alignment between the predicted outputs and ground-truth observations. The AUC values ranged from 0.89 to 0.92, sensitivity varied between 0.89 and 0.90, specificity fell between 0.88 and 0.92, and Kappa coefficients ranged from 0.89 to 0.94. These metrics collectively validate that each model achieved a good to excellent level of predictive agreement, making them reliable tools for analyzing wetland hydrological strength. To determine the most effective model among those tested, a comparative evaluation was carried out. The REPTree algorithm consistently demonstrated the highest performance across all metrics, as shown in Table 5.5. Notably, the MCC for REPTree reached 0.82 for both the pre- and

Table 5.4 Area under different hydrological strength state categories using ML methods

Machine learning algorithms applied	Pre-dam			Post-dam		
	Low	Moderate	High	Low	Moderate	High
ANN	361.74 (33.80)	209.67 (19.59)	498.79 (46.61)	184.38 (33.75)	186.84 (34.20)	174.96 (32.03)
RF	361.96 (33.82)	210.11 (19.63)	498.13 (46.55)	192.24 (35.19)	182.41 (33.40)	171.52 (31.40)
REP_{TREE}	361.89 (33.82)	209.52 (19.58)	498.79 (46.61)	185.8 (34.02)	186.9 (34.22)	173.47 (31.76)

post-dam periods, outperforming the MCC scores of other models.

The spatial results derived from the models revealed that the downstream sections of the wetland experienced notable hydrological instability, often marked by irregular water presence, shortened hydro-periods, and shallower water depths. In contrast, the central wetland zones maintained relatively higher hydrological strength. However, this strength was not uniformly distributed, as some areas proximal to the main river channel were classified as hydrologically degraded, largely due to the presence of embankments that block natural flooding. These areas, often mistaken as non-valuable land, are frequently reclaimed—especially when classified as public lands—without recognition of their ecological function.

To overcome the challenge of limited ground-based hydrological monitoring data, this study employed remote sensing-derived hydrological indicators, including pixel-scale hydro-period, water depth, and WPF. These remotely sensed metrics provide a cost-effective and scalable solution for hydrological modeling across large geographic areas. Nonetheless, future research would benefit from the use of higher-resolution imagery and refined methods for variable extraction. The presence of only one or two field monitoring stations is insufficient to build a spatially robust hydrological strength model, thus reinforcing the importance of image-driven approaches in data-scarce environments. The application of machine learning was instrumental in developing the hydrological strength models, allowing for the modeling of complex spatial relationships among variables while leveraging the computational strengths of these algorithms. This is consistent with the recommendations of Al-Abadi and Al-Najar (2020), Pal and Paul (2020), and Pal and Sarda (2022), who advocate for the use of ML techniques over traditional statistical methods for environmental modeling.

A multi-model approach was adopted in this study to identify the most effective technique, and each model underwent rigorous statistical testing to confirm its suitability. When all evaluation metrics indicate high performance, the

Table 5.5 Accuracy result of the wetland hydrological strength models using ML algorithm

Phases	Models	AUC (ROC)	Sensitivity	Specificity	MCC	Kappa coefficient (K)
Pre-dam	REPTree	0.91	0.9	0.92	0.82	0.91
	RF	0.90	0.89	0.92	0.81	0.90
	ANN	0.91	0.89	0.88	0.78	0.89
Post-dam	REPTree	0.92	0.9	0.92	0.82	0.94
	RF	0.9	0.89	0.92	0.81	0.92
	ANN	0.89	0.89	0.88	0.78	0.92

confidence in that model's reliability is greatly enhanced. In this context, the machine learning models demonstrated a high degree of accuracy and consistency, confirming their utility as essential tools for wetland assessment, restoration, and management.

5.5 Discussion

Since the historical period, these floodplain wetlands have supported widespread ecosystem services (Rudra 2014). However, early settlers and local stakeholders have often overlooked these services and considered these wetlands a source of mosquitoes and the center of emerging epidemics and various health hazards (Hawkes et al. 2020). Such misconception led to the modifications and alterations of wetlands and their surrounding habitat into large-scale farmland initially, and with growing population pressure, these were further transformed into settlement areas with growing infrastructure (Chatterjee et al. 2021; Pal et al. 2022). According to the Space Application Centre (2011) report, India has lost 3% of wetland area every year while 33% of wetland suffered from hydrological instability, among which the majority of the wetland belongs to floodplain area (Sarkar et al. 2020). Apart from direct human intervention, floodplain wetlands are severely affected by flow reduction, decreasing water depth, abnormal hydro-period, and connectivity loss to active recharge points, which increases modification rate and leads to degradation and loss of a valuable environmental asset (Pal et al. 2020; Pal and Debanshi 2021). Many studies like Debanshi and Pal (2020), Paul and Pal (2020), Pal et al. (2020), and Bhattacharjee et al. (2021) actively reported such rapid modifications and alteration of wetland hydrology and habitat in the floodplain regions of north-eastern Bangladesh and lower Atreyee part and Mature and Moribund Deltaic part of India in their studies. Agricultural extension, settlement encroachment, infrastructural development, flow modification, and growing population pressure emerged as the leading factors of such loss and alterations.

References

Ahmed SF, Alam MSB, Hassan M, Rozbu MR, Ishtiak T, Rafa N et al (2023) Deep learning modelling techniques: current progress, applications, advantages, and challenges. Artif Intell Rev 56(11):13521–13617

Al-Abadi AM, Al-Najar NA (2020) Comparative assessment of bivariate, multivariate and machine learning models for mapping flood proneness. Nat Hazards 100(2):461–491

Behnia P, Blais-Stevens A (2018) Landslide susceptibility modelling using the quantitative random forest method along the northern portion of the Yukon Alaska Highway Corridor, Canada. Nat Hazards 90(3):1407–1426

Bertassello LE, Jawitz JW, Aubeneau AF, Botter G, Rao PSC (2019) Stochastic dynamics of wetlandscapes: ecohydrological implications of shifts in hydro-climatic forcing and landscape configuration. Sci Total Environ 694:133765

Bhattacharjee S, Islam MT, Kabir ME, Kabir MM (2021) Land-use and land-cover change detection in a north-eastern wetland ecosystem of Bangladesh using remote sensing and GIS techniques. Earth Syst Environ 5(2):319–340

Borro M, Morandeira N, Salvia M, Minotti P, Perna P, Kandus P (2014) Mapping shallow lakes in a large South American floodplain: a frequency approach on multitemporal Landsat TM/ETM data. J Hydrol 512:39–52

Breiman L (2001) Random forest, vol. 45. In: Mach learn, p 1

Cardinaux F, Enenkl M, Giron F, Kemp T, Uhlich S (2020) U.S. Patent No. 10,564,923. U.S. Patent and Trademark Office, Washington, DC

Chatterjee U, Majumdar S, Bhunia GS (2021) Land restoration strategies to mitigate the problems of shrinkage of wetlands in West Bengal. In: Modern cartography series, vol 10. Academic Press, pp 525–538

Chen W, Hong H, Li S, Shahabi H, Wang Y, Wang X, Ahmad BB (2019) Flood susceptibility modelling using novel hybrid approach of reduced-error pruning trees with bagging and random subspace ensembles. J Hydrol 575:864–873

Chicco D, Jurman G (2020) The advantages of the Matthews correlation coefficient (MCC) over F1 score and accuracy in binary classification evaluation. BMC Genomics 21:1–13

Choi C, Kim J, Han H, Han D, Kim HS (2019) Development of water level prediction models using machine learning in wetlands: a case study of Upo wetland in South Korea. Water 12(1):93

Debanshi S, Pal S (2020) Modelling water richness and habitat suitability of the wetlands and measuring their spatial linkages in mature Ganges delta of India. J Environ Manage 271:110956

Donchyts G, Schellekens J, Winsemius H, Eisemann E, Van de Giesen N (2016) A 30 m resolution surface water mask including estimation of positional and thematic differences using landsat 8, srtm and openstreetmap: a case study in the Murray-Darling Basin, Australia. Remote Sens 8(5):386

Duan G, Niu R (2018) Lake area analysis using exponential smoothing model and long time-series landsat images in Wuhan, China. Sustainability 10(1):149

Hawkes F, Medlock JM, Vaux AG, Cheke R, Gibson G (2020) Wetland mosquito survey handbook: assessing suitability of British wetlands for mosquitoes

Hong Z, Ding S, Zhao Q, Qiu P, Chang J, Peng L et al (2021) Plant trait-environment trends and their conservation implications for riparian wetlands in the Yellow River. Sci Total Environ 767:144867

Khan HM, Rahman MM (2022) Transboundary river management of the Ganges–Brahmaputra–Meghna (GBM) Delta: environmental challenges and strategies. In: Environmental degradation: challenges and strategies for mitigation, pp 495–530

King SL, Laubhan MK, Tashjian P, Vradenburg J, Fredrickson L (2021) Wetland conservation: challenges related to water law and farm policy. Wetlands 41(5):54

Ma F, Chen J, Chen J, Wang T, Han L, Zhang X, Yan J (2021) Evolution of the hydro-ecological environment and its natural and anthropogenic causes during 1985–2019 in the Nenjiang River basin. Sci Total Environ 799:149256

Ma Y, Xu N, Sun J, Wang XH, Yang F, Li S (2019) Estimating water levels and volumes of lakes dated back to the 1980s using Landsat imagery and photon-counting lidar datasets. Remote Sens Environ 232:111287

Mohamed WNHW, Salleh MNM, Omar AH (2012) A comparative study of reduced error pruning method in decision tree algorithms. In: 2012 IEEE International conference on control system, computing and engineering. IEEE, pp 392–397

Mondal D, Pal S (2018) Monitoring dual-season hydrological dynamics of seasonally flooded wetlands in the lower reach of Mayurakshi River, Eastern India. Geocarto Int 33(3):225–239

Mosavi A, Ozturk P, Chau KW (2018) Flood prediction using machine learning models: literature review. Water 10(11):1536

Nampak H, Pradhan B, Abd Manap M (2014) Application of GIS based data driven evidential belief function model to predict groundwater potential zonation. J Hydrol 513:283–300

Nguyen H, Drebenstedt C, Bui XN, Bui DT (2020) Prediction of blast-induced ground vibration in an open-pit mine by a novel hybrid model based on clustering and artificial neural network. Nat Resour Res 29(2):691–709

Orimoloye IR, Kalumba AM, Mazinyo SP, Nel W (2020) Geospatial analysis of wetland dynamics: wetland depletion and biodiversity conservation of Isimangaliso Wetland, South Africa. J King Saud Univ Sci 32(1):90–96

Pal S, Debanshi S (2021) Developing wetland landscape insecurity and hydrological security models and measuring their spatial linkages. Eco Inform 66:101461

Pal S, Paul S (2020) Assessing wetland habitat vulnerability in moribund Ganges delta using bivariate models and machine learning algorithms. Ecol Ind 119:106866

Pal S, Paul S (2021) Stability consistency and trend mapping of seasonally inundated wetlands in Moribund deltaic part of India. Environ Dev Sustain 23:12925–12953

Pal S, Sarda R (2020) Damming effects on the degree of hydrological alteration and stability of wetland in lower Atreyee River basin. Ecol Ind 116:106542

Pal S, Sarda R (2022) Modelling water richness in riparian flood plain wetland using bivariate statistics and machine learning algorithms and figuring out the role of damming. Geocarto Int 37(19):5585–5608

Pal S, Singha P (2022) Image-driven hydrological parameter coupled identification of flood plain wetland conservation and restoration sites. J Environ Manage 318:115602

Pal S, Sarkar R, Saha TK (2022) Exploring the forms of wetland modifications and investigating the causes in lower Atreyee river floodplain area. Eco Inform 67:101494

Pal S, Talukdar S, Ghosh R (2020) Damming effect on habitat quality of riparian corridor. Ecol Ind 114:106300

Paola JD, Schowengerdt RA (1995) A review and analysis of backpropagation neural networks for classification of remotely-sensed multi-spectral imagery. Int J Remote Sens 16(16):3033–3058

Paul S, Pal S (2020) Exploring wetland transformations in moribund deltaic parts of India. Geocarto Int 35(16):1873–1894

Pham BT, Prakash I, Singh SK, Shirzadi A, Shahabi H, Tran TTT, Bui DT (2019) Landslide susceptibility modeling using Reduced Error Pruning Trees and different ensemble techniques: hybrid machine learning approaches. CATENA 175:203–218

Quinlan JR (1987) Decision trees as probabilistic classifiers. In: Proceedings of the fourth international workshop on machine learning. Morgan Kaufmann, pp 31–37

Rudra K (2014) Changing river courses in the western part of the Ganga-Brahmaputra delta. Geomorphology 227:87–100

SAC (Space Application Centre) (2011) National wetland atlas SAC/RESA/AFEG/NWIA/ATLAS/2011

Sarda R, Das P (2018) Monitoring changing trends of water presence state in the major manmade reservoirs of Mayurakshi river basin, eastern India. Spat Inf Res 26:437–447

Sarkar UK, Bakshi S, Lianthuamluaia L, Mishal P, Das Ghosh B, Saha S, Karnatak G (2020) Understanding enviro-climatological impact on fish biodiversity of the tropical floodplain wetlands for their sustainable management. Sustainable Water Resources Management 6:1–12

Shaffer PW, Kentula ME, Gwin SE (1999) Characterization of wetland hydrology using hydrogeomorphic classification. Wetlands 19(3):490–504

Talukdar S, Pal S (2020) Wetland habitat vulnerability of lower Punarbhaba river basin of the uplifted Barind region of Indo-Bangladesh. Geocarto Int 35(8):857–886
Tsoi W, Growns I, Southwell M, Mika S, Lewis S, Ryder D, Frazier P (2022) Effects of inundation on water quality and invertebrates in semiarid floodplain wetlands. Inland Waters 12(3):397–406
Vári Á, Podschun SA, Erős T, Hein T, Pataki B, Iojă IC et al (2022) Freshwater systems and ecosystem services: challenges and chances for cross-fertilization of disciplines. Ambio 51(1):135–151
Warrens MJ, Pratiwi BC (2016) Kappa coefficients for circular classifications. J Classif 33:507–522
World Economic Forum (2019) The global risks report 2019, 14th edn. World Economic Forum, Geneva
Yang Z, Zhou M (2015) Weighted kappa statistic for clustered matched-pair ordinal data. Comput Stat Data Anal 82:1–18
Youssef AM, Pourghasemi HR, Pourtaghi ZS, Al-Katheeri MM (2016) Landslide susceptibility mapping using random forest, boosted regression tree, classification and regression tree, and general linear models and comparison of their performance at Wadi Tayyah Basin, Asir Region, Saudi Arabia. Landslides 13:839–856
Zhao Z, Lou Y, Chen Y, Lin H, Li R, Yu G (2019) Prediction of interfacial interactions related with membrane fouling in a membrane bioreactor based on radial basis function artificial neural network (ANN). Biores Technol 282:262–268

6 Eco-hydrological Modification

Abstract

This chapter innovatively tried to apply water-based indices for exploring the pixel-scale eco-hydrological state including range of variability (RVA)-based flow failure estimation, eco-deficit/surplus analysis, and degree of impact at pixel-scale coupling with historical satellite imageries. Result clearly depicted that in the pre-dam phase, safe area in the post-monsoon season was 37% of the total wetland area, and in post-dam phase, it was decreased to only 9% leading to34% increase of area under high failure rate (FR) (FR > 66%). About 53–62% wetland area was accounted under high eco-hydrological deficit zone in post-dam period. Landscape morphological condition shows that growing fragmentation in post-dam period. Eco-hydrological deficit is more predominant over patch, edge, and small core areas. These statistics prove that damming incident has lead wetland toward increasing hydrological and ecological vulnerabilities. Pixel-scale identification of eco-hydrological stress would be beneficial for regional-level planning and implementation.

Keywords

Degree of hydrological alteration · Ecological optimality · Eco-hydro-surplus/deficit and landscape morphology

6.1 Concept of Eco-hydrological State of the Wetland

Broadly eco-hydrology defines interaction of water with ecosystem including biota, soil, vegetation, atmosphere, etc. (Zalewski et al. 1997). In this chapter, eco-hydrological state of wetland does indicate how water availability, seasonality, variability are ambient for existing hydrological as well as dependent ecosystem. Hydrological system of a wetland strongly determines the species composition, ecosystem structure, biodiversity, even their life cycles (Mitsch et al. 2009). Change in system is also a part of a system; however, changes beyond tolerance often lead to break in system and production failure (Bullock and Acreman et al. 2003). Damming is an external force that primarily determines river flow and subsequently natural floodplain spilling, wetland water, and nutrient supply (Chen et al. 2021). Naiman and Decamps (1997) documented that about 85% of the rivers across the world are tamed by dams of different types and sizes and it certainly leads to changes in flow regime and ecosystem response. Considerable flow attenuation in river may exert acute stress and invite system failure. Hydrological change in a wetland following damming is of different types including (1) shrinkage of active wetland area, (2) increasing inconsistencies in water appear-

S. Pal et al., *Damming and Wetland Eco-hydrology*, GIScience and Geo-environmental Modelling,
https://doi.org/10.1007/978-3-032-11729-8_6

ance, (3) shallowing water depth, (4) narrowing hydro-period, and (5) arrhythmic water availability. When these changes exceed the range of survival, the ecosystem experiences stress, which, if it intensifies, may ultimately lead to ecosystem failure.

For instance, a particular species likes to live in a wetland with at least 2 m, depth but excessive shallowing below 2 m could bring eco-hydrological deficit and pose stress on that species. Similarly, excessive increase of water depth above 2 m could enhance eco-surplus condition in reference to that species which is also not ambient for that species. Similarly change in hydro-period is a stressor on change in biological activities of the wetland biota. Narrowing and lengthening of hydro-period both may exert such pressure. Natural adjustment could help to cope with, and otherwise following natural selection rule, the species could force to migrate other ambient places or be extinct. In this sense, eco-hydrology is a typical state of hydrology which supports the existing ecosystem state and function.

A good many approaches are frequently used for evaluating eco-hydrological condition in post-dam flow condition in river including (1) flow duration curve, (2) low-flow indices, (3) assessment of ecological flow components, (4) range of variability of water flow including magnitude, timing, extremity, duration, etc., (5) hydrological modeling and simulation of ecosystem response, (6) monitoring vegetation and aquatic lives, (7) microscale land cover change, (8) water quality change including physic-chemical and biological ingredients, (9) environmental flow assessment, (10) stakeholders response-based approach and so on. However, most of these approaches are frequently used for analyzing eco-hydrological state of the dam-driven downstream river course. The approaches for evaluating eco-hydrological state (optimum condition) of a wetland in reference to dam are very less attended but it is highly necessary in order to address the ecological changes in parity with hydrological changes (Tonkin et al. 2017; Westwood et al. 2017). In this chapter, water-based indices are tried to be implemented for exploring the eco-hydrological state including range of variability (RVA)-based flow failure estimation, eco-deficit/surplus analysis, and degree of impact at pixel-scale coupling with historical satellite imageries. This innovative approach could help to rightly interpret the eco-hydrological change at pixel scale.

6.2 Materials and Methods

The historical monthly Landsat satellite images as wish to be included could be used for this study. Field-based ground truth could also be used for cross validating the result as used in the present case.

6.2.1 Determination of Eco-hydrological Threshold and Failure Rate

The range of variability approach (RVA) of Richter et al. (1997) is a frequently used flow indices-based approach for determination of eco-hydrological flow condition and its change. In this approach, multiple flow indicators have been taken into consideration and high and low limits are detected. In Chap. 3, detail is described. Different approaches were endorsed by Richter et al. (1997) including (1) 25th and 75th percentile, (2) 20th and 80th percentile, (3) standard deviation (SD) approach (± 1 or $\pm$ 2 SD). In these approaches, individual indicator-specific such methods are applied and high and low threshold are determined. The computed range is called RVA or ecological optima. Indicators value beyond the range signifies failure to attain the ecological optima. At river scale, this application for flow analysis is easier but its application at pixel scale is very difficult. For this, first pixel-scale water depth data of a day/month/ season/year (as required) is essential to compute for a certain period of time and these data could be used for computing threshold values. For this case, SD-based RVA was used considering its wider acceptability. Mean $\pm$ SD1 for each pixel using NDWI images considering as relative water depth was then computed. NDWI images driven from Landsat images from 1990 to 2024 were considered. Based on these, pixel-wise mean and SD were computed. RVA was then defined using Mean ± SD1 formula for each pixel. Entire time spectrum was subdivided into pre- and post-dam period taking damming year (2010) as yard mark. Seasonal eco-hydrological failure rate was then worked out accounting the facts that how frequently (how many years in the present case) wetland water depth values beyond the threshold limits at each pixel (Eq. 6.1). The high failure rate in wetlands does specify eco-hydrological pressure.

$$FR_{EWA} = \left(\frac{\text{Frequency of } NDWI_{FEWA}}{\text{Total frequency of } NDWI_{ADP}}\right) \times 100\%. \quad (6.1)$$

RVA thresholds were computed pursuing the fact that frequency (year) NDWI ($NDWI_{FEWA}$) scores in the post-dam period have failed to reach NDWI ($NDWI_{EWA}$) thresholds or the ecological water state of the wetland.

6.2.2 Degree of the Hydrological Alteration

The degree of hydrological alteration (DHA) demotes the percentage of years of the post-dam period that falls within the threshold range of water depth set based on pre-dam period's water depth. It is computed based on the departure between observed and expected (within the threshold limits) water depth considering time series NDWI. The depth of wetland water in the pre-dam condition is considered as expected depth (ED) and in the post-dam condition is considered as observed depth (OD). The logic behind this consideration is that if dam was not installed or the flow was not controlled, water depth was as it is as was during pre-dam period. Based on this information, degree of hydrological alteration is computed following Eq. 6.2.

$$\text{DHA } (\%) = \left(\frac{\text{OD} - \text{ED}}{\text{ED}}\right) \times 100. \quad (6.2)$$

The degree of alteration would be positive, negative, or zero. Negative alteration denotes that observed values fall within the threshold of eco-hydrological limits less often than expected. A positive change contrarily shows that the observed value falls more often than the expected values within the eco-hydrological threshold limits and zero value means in the post-dam condition the observed water depth is almost equivalent with expected water depth.

6.2.3 Eco-hydro-surplus and Eco-hydro-deficit for Examining Ecological Habitat Suitability of Wetlands

Eco-surplus and deficit are usually computed considering the flow duration curve (FDC). Longer data series with higher frequency is good. For doing this, FDCs of unregulated and regulated flow are first constructed and compared. Here regulated means dam-driven flow and unregulated means pre-dam flow condition. Particularly in reference to wetland water depth, regulated water depth is referred to as water depth of post-dam period and unregulated water depth is meant by water depth during pre-dam period.

Eco-surplus is characterized while the regulated (post-dam) median annual flow (water depth here) duration curve (FDC) remains above the unregulated (pre-dam) FDC. Conversely, eco-deficit is the position of the regulated median annual FDC below the unregulated median yearly FDC (Vogel and Fennessey 1995). This technique helps to examine the overall gain or loss of water flow in a river in post-dam period (Vogel et al. 2007). In this study, this approach is chosen for measuring the altered water availability state of wetlands. FDC is determined using mean NDWI scores of a considered period. Based on the finding of this pixel-wise eco-deficit and surplus mapping was also done.

6.2.4 Landscape Morphological Change of Wetland

Landscape morphology is itself a good indicator of wetland hydrological dynamics and land use dynamics. The perennial and consistent wetland is less susceptible to landscape morphological transformation than seasonal, ephemeral wetlands that are hydrological irregular, erratic, and provide a scope of activating invasive factors of wetland morphological change (Lee et al. 2020). Applying Fragstat software, wetland landscape of both pre- and post-dam periods was categorized into patch, edge, perforated, small, medium, and large core. Area under each category was also accounted in order to highlight the morphological changes of the wetlands. Changein patch frequency, area, edge area, and a core area in between pre- and post-dam periods could help to explicate the nature of influencing factors.

6.3 Results

6.3.1 Hydrological Threshold and Failure Rate Assessment

From the outcome of the hydrological state models as applied earlier of this work, it is very clear that damming has exerted immense consequences on bringing a hydrological change to the wetland. Work of this chapter further tried to examine whether this change has exceed the eco-hydrological thresholds. Apart from this, spatial pattern of eco-hydrological state could help to restoration and conservation of wetland at regional scale with priority basis.

If, at a wetland pixel, the water depth remained within the defined threshold limits, it was considered eco-hydrologically safe. To illustrate spatial differences more clearly, the degree of failure rate recorded at each pixel was further categorized, as shown in Fig. 6.1. In the pre-dam phase, the proportion of wetland area falling within the safe zone during the post-monsoon season was 37%, which decreased sharply to 9% in the post-dam phase. In the pre-monsoon season as well, the extent of the safe zone was drastically reduced. Conversely,

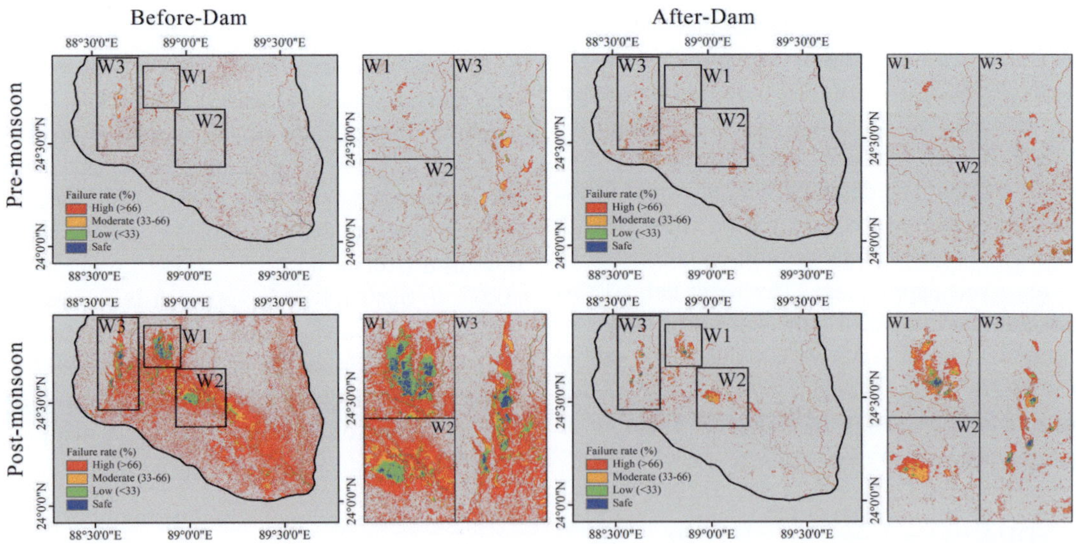

Fig. 6.1 Condition of spatio-seasonal (pre- and post-monsoon seasons) failure rate of wetland water depth in pre- and post-dam states

the area under zones with high failure rates (FR > 66%) increased by 34% and 7% in the pre- and post-monsoon seasons, respectively. Similarly, the area under zones with moderate failure rates (FR between 33 and 66%) increased by 6 and 14% for the pre- and post-monsoon seasons (Table 6.1). In many instances, the FR values fell below the lower threshold limit, indicating increasing eco-hydrological susceptibility of the wetlands.

6.3.2 Degree of the Eco-hydrological Impact/Alteration

Figure 6.2 demonstrates the degree of eco-hydrological alteration triggered by damming. The impact at spatial scale is subdivided into six classes in order to explore the spatial differentiation. Eco-hydrological impact classes are (1) un-impacted moribund, (2) low negative impact < 33%, (3) high negative impact > 33%, (4) low positive impact < 33%, and (5) high positive impact > 33%, and (6) un-impacted (safe) un-impacted moribund wetlands are recognized those which were already desiccated or reclaimed. About 71% of the wetland area during post-monsoon season was accounted un-impacted moribund in the post-dam period. Remaining 29% of the wetland area recorded negative impact. A negative degree of impact signifies that the expected water depth is less than the observed water depth. Out of 29%, about 20% and 9% of wetland areas were recognized as a negative degree of impact in reference to < 33% and > 33%, respectively (Table 6.2). No area witnessed a positive change which is not ecologically viable. It reveals the fact that dam-driven hydrological alteration in river adversely transformed the eco-hydrological state of the wetland.

6.3.3 Eco-hydro-surplus and Eco-hydro-deficit of Wetland

Eco-hydro-surplus and eco-hydro-deficit of wetland area in reference to water depth showed pixel-scale eco-hydro-deficit/surplus based on the FDC of regulated and unregulated flow conditions in relation to wetland water depth (Fig. 6.3). The FDC in both seasons clearly demonstrated eco-hydro-deficit conditions. No eco-hydro-surplus condition was recorded at any exceedance probability. Spatio-seasonal analysis at pixel scale confirmed the complete eco-deficit condition. Not a single pixel, even at the core wetland parts, was recognized as possessing eco-hydro-surplus. For better representation and analysis, the eco-hydro-deficit zone was further categorized into three subclasses based on their degree of deficit. These zones are: (1) deficit level 1 (< 33%), (2) deficit level 2 (33–67%), and (3) deficit level 3 (> 67%). Area statistics of the concerned zones showed that areal concentration was substantially higher in the greater deficit zones in both pre- and post-monsoon seasons (Table 6.3).

Outskirt wetland areas and smaller wetlands were found highly susceptible to eco-hydro-deficit, and even some core wetland parts close to the main river also experienced higher-level eco-hydro-deficit states. In the pre-monsoon season, due to lack of rainfall and high temperature, except for the main river and a few deeper wetland patches, most parts of the wetland dried out. This indicates growing ecological stress in the downstream riparian wetlands (Fig. 6.3).

The case sites were taken from different levels of eco-hydro-deficit zones. The degree of hydro-ecological alteration of each wetland under different eco-hydro-deficit zones varied accordingly. Wetlands located in the high eco-hydro-deficit zone (Raktadaha, Chatra, and Chalan Beel) witnessed stronger hydrological transformation and ecological diminution. On the contrary, wetlands nearer to the parent river at low eco-hydro-deficit zones (Halti Beel and Bhabanipur) also experienced negative eco-hydrological changes, but at a much lower rate (Fig. 6.3). The lowering of discharge, flood frequency, flood spilling extent, and hydrological connectivity between river and wetland across a wide part of the study region caused eco-hydrological alteration of wetlands. Hydrologically isolated wetlands and those located farther from the main rivers

Table 6.1 Areal extent under safe and different failure rate zones

Rate of failure (%)	Pre-monsoon				Post-monsoon			
	Before dam		After dam		Before dam		After dam	
	Area in km^2	% of area	Area in km^2	% of area	Area in km^2	% of area	Area in km^2	% of area
Low	21.53	47.09	2.78	13.18	87.49	10.04	73.62	17.12
Moderate	12.07	26.39	6.82	32.33	108.06	12.4	114.42	26.61
High	9.15	20.01	11.50	54.49	351.90	40.38	202.62	47.12
Safe	2.98	6.51	0.00	0	324.01	37.18	39.35	9.15
Total	45.72	100	21.10	100	871.46	100	430	100

witnessed the greatest transformation. To reduce this growing vulnerability, Nikitina et al. (2020) and Hoque et al. (2022) strongly recommended the release of ecological flows.

6.3.4 Transformation of Landscape Morphological Scenario of the Wetland

Figure 6.4 depicts the landscape morphological state of the wetland in the pre- and post-dam periods. Result clearly reported that most of the large core wetland reduced their size due to insufficient water supply. In the pre-dam period, a long wetland tract was mapped on both the sides of the Atreyee River. In post-dam period, the scenario of wetland landscape was changed significantly. Most of the continuous large wetland corridor was found fragmented. For example, in the pre-dam period, 11.79 and 23.21% of areas were reported patch and edge respectively but these were significantly increased to 28.2% and 32.74% in same order during the post-dam period. Perversely, the large core area was squeezed from 47.52 to 16.75% showing 30.77% reduction in between pre- and post-dam periods (Table 6.4). Growing fragmentation implies intensifying the patch and edge area which promotes the augmentation of anthropogenic intrusion (Shen et al. 2019).

6.4 Discussion

Riparian wetlands in subtropical region are majorly fed by seasonal rainfall regime and river water. Seasonal rainfall regime is strongly responsible for strengthening and weakening of hydrological condition of the wetland. During monsoon season, due to having maximum rainfall (about 80%), in monsoon and post-monsoon seasons, wetlands get hydrologically energized.

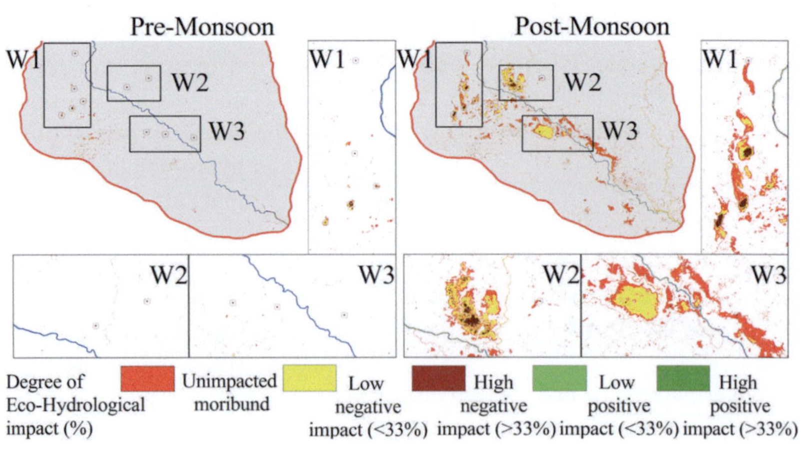

Fig. 6.2 Spatio-seasonal scenario of the degree of eco-hydrological impact in after damming phase

Table 6.2 Seasonal proportion of area under varying eco-hydrological impacts

Degree of impact	Pre-monsoon (km^2)	Area in %	Post-monsoon (km^2)	Area in %
Un-impacted moribund	16.80	79.63	304.66	70.85
Low negative impact (< 33%)	3.97	18.8	84.58	19.67
High negative impact (> 33%)	0.33	1.54	40.76	9.48
Low positive impact (< 33%)	0.01	0.02	0	0
High positive impact (> 33%)	0	0	0	0
Un-impacted (safe)	0	0	0	0

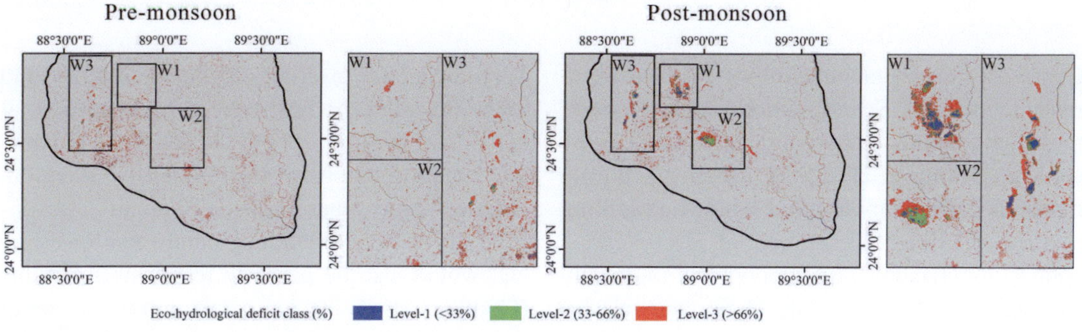

Fig. 6.3 Eco-hydro-deficit/surplus in the post-dam period in both the seasons (left and right-hand side figures above), FDCs of regulated and unregulated conditions of wetland water depth in pre- and post-monsoon seasons (FDCs inserted below)

Table 6.3 Areal extent of wetland under different eco-hydro-deficit levels in both the seasons

Eco-hydro-deficit class	Pre-monsoon		Post-monsoon	
	Area in km^2	% of area	Area in km^2	% of area
Level 1 (< 33%)	7.65	36.24	83.72	19.47
Level 2 (33–67%)	2.29	10.84	78.65	18.29
Level 3 (> 67%)	11.17	52.92	267.59	62.23
Total	21.10	100	430	100

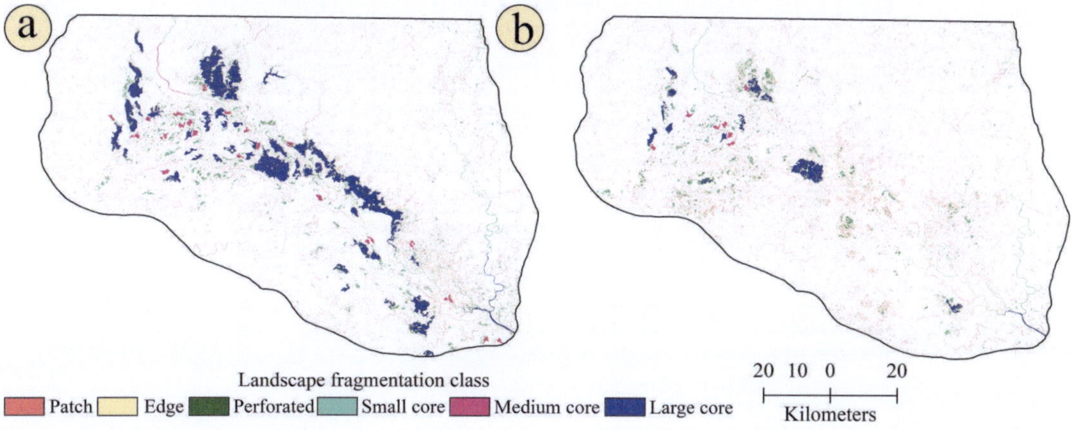

Fig. 6.4 Wetland landscape fragmentation depicting patch, edge, perforated, small, medium, and large cores during **a** before dam and **b** after dam periods

Table 6.4 Area under different landscape fragmentation units computed in reference to post-monsoon season of before and after damming conditions

Classes	Pre-dam		Post-dam		Change between pre-dam to post-dam	
	km^2	%	km^2	%	km^2	%
Patch	102.75	11.79	121.26	28.2	18.51	16.41
Edge	202.27	23.21	140.78	32.74	– 61.48	9.53
Perforated	20.48	2.35	17.93	4.17	– 2.55	1.82
Small core	99.70	11.44	69.49	16.16	– 30.21	4.72
Medium core	32.16	3.69	8.51	1.98	– 23.64	– 1.71
Large core	414.12	47.52	72.03	16.75	– 342.09	– 30.77

Contrarily, reverse picture is observed during pre-monsoon season. Scanty rainfall, high temperature, high evaporation hydrologically weakens the wetland (Xiong et al. 2023). Moreover, during monsoon and post-monsoon season, flood spilling in concerned river, activation of river-wetland connecting channels also support hydrologically strengthening wetland. Role of tie channels for making wetland hydrologically rich was described by Pal et al. (2023). This is the scenario of seasonal hydrological regime of wetland. This clearly depicts hydrological deficit during pre-monsoon season and surplus during monsoon season. But this natural hydrological regime was also noticeably affected by damming over the river in 2010. Gradual diminishing of wetland condition is well reported even in the unregulated river basin but this changing rate has become alarming after dam-driven diversion of river water. Very high rate lowering of water depth beyond lower RVA threshold, sharp eco-hydro-deficit all across the wetland in both the seasons are some good manifestations of this human intervention. Besides these other forms of human interventions like water extraction, groundwater lowering, etc. would also be partially responsible for this (Gumiero et al. 2013).The dwindling of water delivery to the wetland made an eco-hydrological crisis within the wetland leading to the mounting FR across wider parts of the wetland in aim to attain the ecologically relevant threshold water depths in the present study. It led to the escalating eco-hydrological impact of hydrological alteration. This also insisted the growing eco-hydro-deficit of the existing wetland. Kundu et al. (2021) reported this could reduce the ecological serviceability of the wetland. In order to present the situation in reference to some specific wetlands,

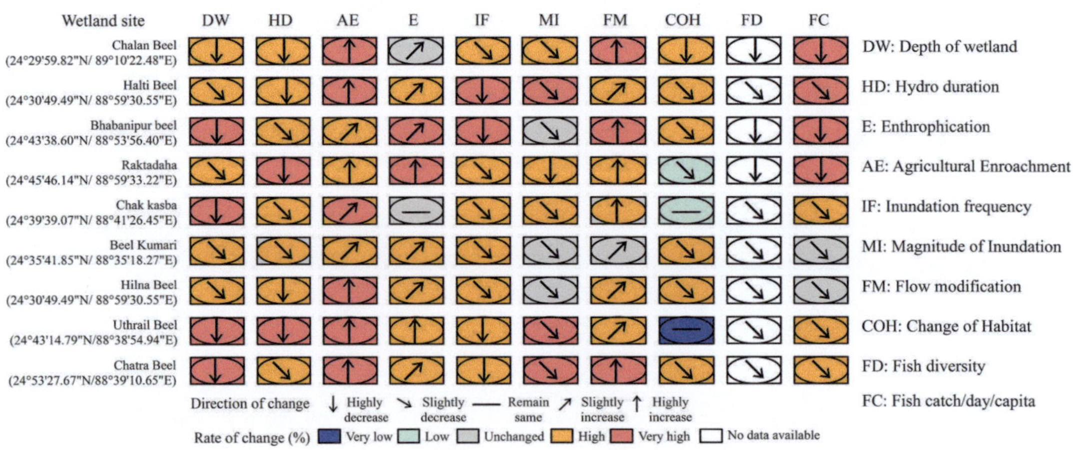

Fig. 6.5 Hydrological and ecological transformation and associated drivers in reference to stakeholders' view of different surveyed wetlands located within different eco-hydrological zones

nine wetland cases were chosen and accounted different hydrological conditions in relation to the associated causes. Irrespective of their location, shape, size, origin, there was found growing dwindling of hydrological condition in post-dam scenario with varying rates as reported by the local stakeholders. Among the causes of changes, stakeholders strongly agreed that damming over river Atreyee is one of the main reasons behind hydrological diminution. The qualitative answer of the stakeholders was presented in Fig. 6.5.

This finding supports the finding received from image-driven analysis. Although computation of eco-deficit/surplus in dam-driven river flow is very common (Wang et al. 2017), such pixel-scale computation of eco-hydro-deficit/surplus is very less but effective in order to obtain spatial-level planning directory.

References

Bullock A, Acreman M (2003) The role of wetlands in the hydrological cycle. Hydrol Earth Syst Sci 7(3):358–389

Chen L, Wu Y, Xu YJ, Zhang G (2021) Alteration of flood pulses by damming the Nenjiang River, China-Implication for the need to identify a hydrograph-based inundation threshold for protecting floodplain wetlands. Ecol Ind 124:107406

Gumiero B, Mant J, Hein T, Elso J, Boz B (2013) Linking the restoration of rivers and riparian zones/wetlands in Europe: sharing knowledge through case studies. Ecol Eng 56:36–50

Hoque MM, Islam A, Ghosh S (2022) Environmental flow in the context of dams and development with special reference to the Damodar Valley Project, India: a review. Sustain Water Resour Manage 8(3):62

Kundu S, Pal S, Talukdar S, Mandal I (2021) Impact of wetland fragmentation due to damming on the linkages between water richness and ecosystem services. Environ Sci Pollut Res 28(36):50266–50285

Lee O, Kim HS, Kim S (2020) Hydrological simple water balance modeling for increasing geographically isolated doline wetland functions and its application to climate change. Ecol Eng 149:105812

Mitsch WJ, Gosselink JG, Zhang L, Anderson CJ (2009) Wetland ecosystems. Wiley

Naiman RJ, Decamps H (1997) The ecology of interfaces: riparian zones. Annu Rev Ecol Syst 28(1):621–658

Nikitina OI, Dubinina VG, Bolgov MV, Parilov MP, Parilova TA (2020) Environmental flow releases for wetland biodiversity conservation in the Amur River Basin. Water 12(10):2812

Pal S, Chowdhury P, Singha P, Let M (2023) Role of tie channel on wetland hydrological security and sustenance. J Clean Prod 418:138162

Richter B, Baumgartner J, Wigington R, Braun D (1997) How much water does a river need? Freshw Biol 37(1):231–249

Shen G, Yang X, Jin Y, Xu B, Zhou Q (2019) Remote sensing and evaluation of the wetland ecological degradation process of the Zoige Plateau Wetland in China. Ecol Ind 104:48–58

Tonkin Z, Kitchingman A, Lyon J, Kearns J, Hackett G, O'Mahony J et al (2017) Flow magnitude and variability influence growth of two freshwater fish species in a large regulated floodplain river. Hydrobiologia 797:289–301

Vogel RM, Fennessey NM (1995) Flow duration curves II: a review of applications in water resources planning 1. JAWRA J Am Water Resour Assoc 31(6):1029–1039

Vogel RM, Sieber J, Archfield SA, Smith MP, Apse CD, Huber-Lee A (2007) Relations among storage, yield, and instream flow. Water Resour Res 43(5)

Wang Y, Wang D, Lewis QW, Wu J, Huang F (2017) A framework to assess the cumulative impacts of dams on hydrological regime: a case study of the Yangtze River. Hydrol Process 31(17):3045–3055

Westwood CG, England J, Dunbar MJ, Holmes NT, Leeming DJ, Hammond D (2017) An approach to setting ecological flow thresholds for southern English chalk streams. Water Environ J 31(4):528–536

Xiong Y, Mo S, Wu H, Qu X, Liu Y, Zhou L (2023) Influence of human activities and climate change on wetland landscape pattern—a review. Sci Total Environ 879:163112

Zalewski M, Janauer GA, Jolánkai G (1997) Ecohydrology: a new paradigm for the sustainable use of aquatic resources

7 Lateral Hydrological Disconnection and Wetland Resilience

Abstract

Regular flood spilling and water, nutrient supply from river to floodplain wetland continue the eco-hydrological vibration of floodplain ecosystem. Whether damming is anyway caused for interruption in this river-wetland connection is a matter of concern and this chapter tried to explore this. For this discharge-based 2D inundation, simulation in HEC-RAS was done in case of pre- as well as post-dam. Overlaying of wetland with regards to the inundation extent in both the hydrological regimes was done in order to classify wetland's eco-hydrological stress. Result revealed that in post-dam period, inundation area was decreased 4828–1627.30 km^2. Inundation depth was also found significantly declined in parity with flow reduction (66% of peak flow decreased) in Master stream. About 81 km^2 (18.8%), 95 km^2 (22.1%), and 338 km^2 (78.6%) of identified wetland areas of the post-dam period were recognized respectively as critical, stress, and safe wetland security state (WSS). Majority of smaller wetlands located at the elevated outskirts zones are critical in nature due their location outside the lateral flood limit of pre-dam period. Often these are ortho-fluvial in nature. Comparison between hydrological stability and WSS further revealed that wider part of low hydrologically stable wetland falls under critical and stress WSS exhibiting the impact of squeezing lateral flood extent.

Keywords

River spilling · Floodplain wetland · Water supply · Nutrient support · Squeeze of lateral flood spread · And Eco-hydrological stress

7.1 Forms and Functions of Lateral Connectivity

Lateral connection between river and wetland is very vital for the survival of floodplain wetland. River water is one of the major water and nutrient sources of wetland. In subhumid region, river spilling mechanism during monsoon season supplies huge water to the floodplain and wetland. This is the basic of floodplain genesis and sustenance (Middleton 1999). Hence, the frequency, magnitude, duration of river spilling play very decisive role toward the survival of wetland hydrology and ecology (Casanova and Brock 2000). Flood spill water carries sediment to the floodplain and renew floodplain, replenish wetland water, recharge groundwater, supplies various forms of nutrients essential for aquatic ecosystem, disperse pawns and other creatures, washes out floodplain pollutants, and so on (Freitag et al. 2012). But climate change and any anthropogenic interventions can reduce flow in river affecting the spilling mechanisms.

S. Pal et al., *Damming and Wetland Eco-hydrology*, GIScience and Geo-environmental Modelling,
https://doi.org/10.1007/978-3-032-11729-8_7

In Chap. 2, dam-driven attenuation of flood frequency, magnitude is well noted in reference to the Atreyee River. Chen et al. (2021) also reported similar incidents in the Nenjiang River, China. This incident is not merely a reduction of river flow but it also poses serious adversity on floodplain processes, functions, and ecological ambience.

Beside, spilling, seepage water-driven, tie channel-driven lateral connections are also very crucial for floodplain wetland sustainability. Seepage water from river hydrologically supports the river proximate wetlands. Tie channels connect the river and wetland and it carries flow to the wetland. So, if the river is not even spilled, water influxes from river to wetland along with substantial amount of sediment, nutrient, fishes, etc. (Mitsch 1992). The amount and duration of flow are strongly dependent on flow dynamics of the river. If flow level of river remains above the bed level of tie channel, water flows through this to the wetland. In natural situation, always these channels are not hydrologically active. Damming may further reduce the flow availability of tie channels and that may enhance water scarcity in dependent wetlands. Pal and Let (2024) showed that deactivation of flow not only affects the wetland functions but also morphological dimensions of both tie channel itself and wetland.

Within floodplain, the hydro-ecological situations are not uniform. For instance, the wetlands within the regularly flood spilling area are not same as those are located beyond the regularly spilled area. Similarly, the wetland which is well connected with hydrologically active tie channels is not vulnerable than those are isolated. Wetlands located at very proximity to river with sufficient seepage flow are hydrologically more stable than those are located away from river. So, all these conditions collectively help to classify wetlands into two distinct classes: (1) para-fluvial and (2) ortho-fluvial. Para-fluvial wetlands are hydrologically well connected with the river and used to receive regular water. Hydrological dynamics of river actively controls the dynamics of wetland. Along river Ganga, Kalindr, Pagla, Adi Ganga, Jalangi in Bengal basin, such forms of wetlands are common. Contrarily, ortho-fluvial wetland receives very infrequent water from river or sometimes entirely isolated from river and thereby experiences hydro-ecological stress (Ward 1989). The evidences of transformation of para-fluvial wetland to ortho-fluvial are also very common in Bengal basin. The hydrological modification in floodplain may bring transformation of wetland state. For instance, if damming reduces flood spilling incidents, tie-channel functions; it may transform many para-fluvial wetland into ortho-fluvial wetland. Flood spill area demarcation in different hydrological regimes could help to classify wetland with their level of hydro-ecological stress. This chapter tried to explore how flow reduction in river is caused for change in active flood extent in post-dam conditions and how these are caused for bringing ecological stress on wetland.

7.2 Materials and Methods

Three hourly discharge data taken from Atreyee River's gauge stations and discharge measured in some tributary rivers using area-velocity method in 2019 to 2020 were taken for lateral flood simulation. People's responses regarding flood evidences and water depth during flood were also digitized for validating simulated flood map. Digital elevation model was used as a base data over which flow simulation was done. Wetland maps developed from Landsat images in pre- as well as post-dam periods were used for further analysis.

7.2.1 Methods for Flood Simulation

Flood simulation explains and maps how a certain flow in a river could create inundation surface in a river basin or a selected area. Simulated spatial extent and water depth depend on amount of discharge, duration of discharge and topographic configurations. It could be done following one or two or three dimensional (1D or 2D or 3D) approaches.

The 1D model is used for prismatic channel (channel having constant cross section). While for non-prismatic (having changing cross section area, configuration) for instance river, floodplain 2D model is implemented (Al Amin and Haki 2017). For 2D simulation of flood inundated area, HEC-RAS version 5.0, frequently applied. The model of 3D hydraulic identifies the properties of flow which is having three directions viz, longitudinal (direction of downstream through x alliance) transversal (through y alliance), and vertical (through z alliance). 3D models are quite uncommon in hydraulic studies due to high computational requirements. 2D approach is more reasonable since its manageable computational requirements (Shen and Diplas 2008). It helps to simulate flood with its spatial sprawl and vertical depth. The new HEC-RAS-v5 is capable to solve both 2D diffusive wave and full 2D Saint Venant equations. In this study, 2D inundation was simulated using the depth-averaged Saint-Venant (shallow-water) equations (Eqs. 7.1–7.3).

$$\frac{\partial s}{\partial t}+\frac{\partial x}{\partial p}+\frac{\partial y}{\partial q}=0 \tag{7.1}$$

$$\frac{\partial x}{\partial t}+\frac{\partial}{\partial p}\left(\frac{x^2}{h}\right)+\frac{\partial}{\partial q}\left(\frac{xy}{h}\right) = -\frac{n^2xg\sqrt{x^2+y^2}}{h^2}-gh\frac{\partial s}{\partial p}+xf + \frac{\partial}{\omega\partial p}(h\tau_{pp})+\frac{\partial}{\omega\partial q}(h\tau_{pq}) \tag{7.2}$$

$$\frac{\partial y}{\partial t}+\frac{\partial}{\partial q}\left(\frac{y^2}{h}\right)+\frac{\partial}{\partial p}\left(\frac{xy}{h}\right) = -\frac{n^2yg\sqrt{x^2+y^2}}{h^2}-gh\frac{\partial s}{\partial q}+yf + \frac{\partial}{\omega\partial q}(h\tau_{qq})+\frac{\partial}{\omega\partial p}(h\tau_{pq}), \tag{7.3}$$

where h signifies water depth in (m), x and y denote the flow direction in the direction of p, q ($m^2\ s^{-1}$), surface elevation denoted by s (m), resistance of manning is denoted by n, g marked as how acceleration took place due to gravity ($m\ s^{-2}$), water density denoted by ω ($kg\ m^{-3}$), τ_{pp}, τ_{qq} and τ_{pq} demarcated as stress of effective component of shear. Coriolis is denoted by f (s^{-1}). Equation 7.2 and 7.3 missed out in case of favor of equation of momentum in diffusive wave (Quirogaa et al. 2016).

In case of 2D simulation of flood, maximum, median flow of a selected period (say pre- or post-dam periods) could be taken. Flood simulation also could be done based of flow probability analysis, specifically at different flood probability levels. Following median-based analysis, flow conditions of 1984 and 2014 were selected for representing pre- and post-dam flood situations. After predicting flood probability, flood simulation could also be done for coming years. For details, one may further consult Pal and Talukdar (2018).

7.2.2 Measuring Hydrological Strength

For both the time periods, satellite imageries used to show the hydrological state including water presence frequency (WPF), water depth (WD), and hydro-period of wetland. These were integrated using machine learning (RF, ANN, and REPtree) algorithms. The best representative model among was taken for analyzing how different hydrological strength categories (high, moderate, and low as done here) get support from inundation. Inundation also helps to improve the hydrological strength. The detail of developing hydrological strength has been explained in Chap. 5.

7.2.3 Inundation-Based Classification of Wetland Security State (WSS)

Lateral inundation zones of both dam periods were overlapped and categorized into three inundation zones, namely (1) areas inundated in pre-dam period and are getting inundated in post-dam time period (active flood zone), (2) areas which are situated beyond post-dam lateral inundation limit but inside the inundation limit of pre-dam period, (3) areas fall beyond the lateral inundation limits of the pre- and post-dam periods. Wetlands located in the defined three inundation zones were in that order considered as (1) safe condition, (2) stress condition, and (3) critical condition of hydrological security. Since the wetland receives inundation water due to having its location within active floodplain, the wetland could be treated as safe. Conversely, if the wetland located beyond the flood limits of pre- and post-dam periods, wetlands had been witnessing long term dearth of inundation water scarcity and therefore, these could be treated as stress. These could only survive with the aid of rain water and seepage water (if any). This classification could not only present the lateral inundation connectivity of wetland but also provides insight of nutrient supply to the para-fluvial wetland. It is ecologically very important.

7.2.4 Wetland Hydrological Security and Strength Matrix

Hydrological strength (HS) and wetland hydrological security state (WSS) matrix, an innovative approach, could be applied to investigate how far the hydrological strength is associated with security state of wetland. It was stated earlier that WSS was classified into safe, stress, and critical conditions. On the other hand, hydrological strength was categorized into low, moderate, and high. Since both WSS and HS maps have three subcategories, total nine WSS-HS zones would hypothetically be created depicting likelihood of varying WSS and HS conditions including stress WSS and low HS, safe WSS and high HS, critical WSS and low HS, stress WSS and moderate HS, etc. Each of the subdivision offers a good combination of clear attributes of both. Higher association between low HS under stress and censorious state of WSS and high HS and safe WSS is crucially important for depicting the fact that flood ambience is a paramount factor behind hydrological stability of floodplain wetland.

7.3 Results

7.3.1 Changes in Inundation Character

Inundation zone of pre- as well as post-dam flood spell discharge of the median flood years were demarcated following 2D simulation approach and presented in Fig. 7.1. It is obvious that inundation area was high during pre-dam than post-dam period. Inundation area was about 4828 km^2 during pre-dam period and the same was reduced to 1627.30 km^2 in post-dam period. Depth of inundation was also found significantly higher in pre-dam period (Fig. 7.1). This is the direct consequence of damming toward dwindling of water flow in Master and major tributaries. In Chap. 2, it has clearly been demonstrated that flow volume during monsoon season was declined by 66%. Except 2017, in post-dam period, no year was detected when observed water level crossed extreme danger level leading to high flood. Such

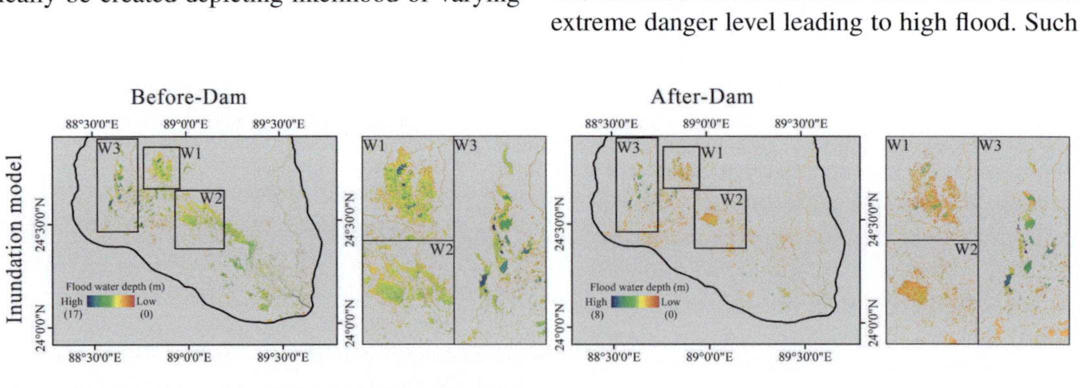

Fig. 7.1 Inundation models in before and after damming periods

a huge change in river flow regime is one of the major causes behind change in inundation extent and water depth.

7.3.2 Wetland Hydrological Security State (WSS) in Connection with Inundation

Figure 7.2 portrays the wetland security states of post-dam (safe, stress, and critical) identified based on their location in regards to active flood area of both dam periods. About 81 km^2 (18.8%), 95 km^2 (22.1%), and 338 km^2 (78.6%) of identified wetland of the post-dam period were recognized respectively as critical, stress, and safe wetland security states. Majority of smaller wetlands located at the elevated outskirts zones are critical in nature due their location outside the lateral flood limit of pre-dam period. Often these are ortho-fluvial in nature. Such isolated wetlands do not receive any river water through inundation, seepage, or tie channels. Larger wetlands at very proximity to main channel are relatively safe in nature due having direct flood water connection. Some stress prone wetland patches were identified at very river proximate area due to insufficient supply of flood water reaching there.

7.3.3 Hydrological Strength

Chapter 5 provides relevant information about the strength of hydrological condition of wetlands in the Atreyee River basin.

7.3.4 Hydrological Strength/Resilience Scenario of Various Hydrological Security Zone (Based on WSS, HS Matrix)

Does lateral connectivity through flooding support hydrological strength and resilience? In order to resolve this issue, spatial-level contingency matrix between WSS and HS was done. It was tried to identify how much wetland extent with higher HS is associated with safe WSS zone. Good association between them could help to approximate that the higher HS supports greater wetland security and otherwise. Figure 7.2 portrays the spatial form of HS and WSS matrix for post-dam period. From these, precisely it could be inferred that spatially no such strong relationship was found between them. Some wetlands were found safe in terms of WSS but always they are not hydrologically in good state. An area covering 90 km^2 wetland was classified under high HS and safe WSS zones, 11 km^2 was identified under critical WSS and high HS zones, 71 km^2 was under moderate HS and critical WSS zones. All these facts clarified that the interrelationship between WSS and HS is not very straight forward. Greater concentration of wetland area under safe WSS and high HS was found because spatial distribution of wetland is more alongside main river and flood water support is more certain on those areas (Table 7.1).

Table 7.1 Hydrological strength condition of various hydrological security state zones

WSS category	HS condition	Area after damming period	
		Area in km^2	Area in %
Safe	High	89.87	20.9
	Moderate	54.05	12.57
	Low	43.77	10.18
Stress	High	16.68	3.88
	Moderate	41.19	9.58
	Low	33.33	7.75
Critical	High	10.88	2.53
	Moderate	70.61	16.42
	Low	69.62	16.19

7.4 Discussion

Shrinkage of inundation area was identified in post-dam period leading to keep many wetlands beyond present-day flood limit. It is one of the causes behind conversion of well fluvial flood connected wetland into disconnected wetland. Huge discharge curtailing through damming is one of the major causes behind such shrinking flood spill area. Tockner et al. (1999), dos Santos Pinto et al. (2020) reported that the wetlands at very proximity to main river with good hydrological connectivity with river are called as ***para-fluvial*** wetland. Regular water, sediment, nutrient exchange between river and wetland helps to distinguish para-fluvial wetland from ortho-fluvial wetland. Conversely, wetlands those are located away from river and are not hydrologically connected could be defined as ortho-fluvial wetland. In the present case, identification of safe, stress, and critical wetland in reference to present and past inundation areas is a new dimension of this work. Certainly, there is much scope to refine this approach. However, the present form of wetland classification in reference to pre- and post-dam inundation extents could also be followed. In the present study, outskirt wetlands and wetland located over relatively elevated floodplain were crucially recognized under stress and critical zones in post-dam period. Curtailing lateral inundation connectivity is a major reason behind this.

To make a floodplain wetland hydrologically and ecologically sound, regular water supply by flooding is one of the prerequisite. For example, it helps to grow nutrient, fish without any cost. It reduces fishing cost which in turn supports the fishermen (Sarkar et al. 2020). One of the causes of ecological health deterioration of wetland is presence of pesticide's particles (Quintela et al. 2019) which help to grow the species which are exotic in nature and on the other hand decreased the species growth which is indigenous in nature (Maurya et al. 2019). Flooding in a regular basis helps to rejuvenate the habitat health condition of wetland (Liu et al. 2018). Based on lateral flooding extent, the wetlands were devised into safe, stressful, and critical condition. The wetland which is critical in nature is not taking the floodwater for both the hydrological regimes. Its presence or absence fully depends on the availability of rainwater seepage. The situation like, critical WSS took place when rainfall became inconsistent and table of groundwater lowered down. The stress WSS zone is characterized by rarely occurrence of flood water. So, these wetlands denoted the poor eco-hydrological situation and poor resilience. Safe zone obtained flood water and nutrient in a regular basis indicating a good eco-hydro state. Talukdar and Pal (2020) already did these types of characterization.

HS-WSS matrix provided enough insight regarding the association between hydrological dynamics, lateral flood dynamics, and resilience of wetland. Some cases at very proximity to the river were found where the association between hydrological strength and wetland security state is good exploring the fact that along with other factors flood plays vital role to control the hydrological condition of these wetlands. Some patches of wetland were found where inundation connectivity is good but hydrological strength is poor. This picture apparently seems contrasting but the fact is that hydrological dynamics is not solely controlled by flood dynamics (Bornette and Amoros 1996). Seepage character, topographical

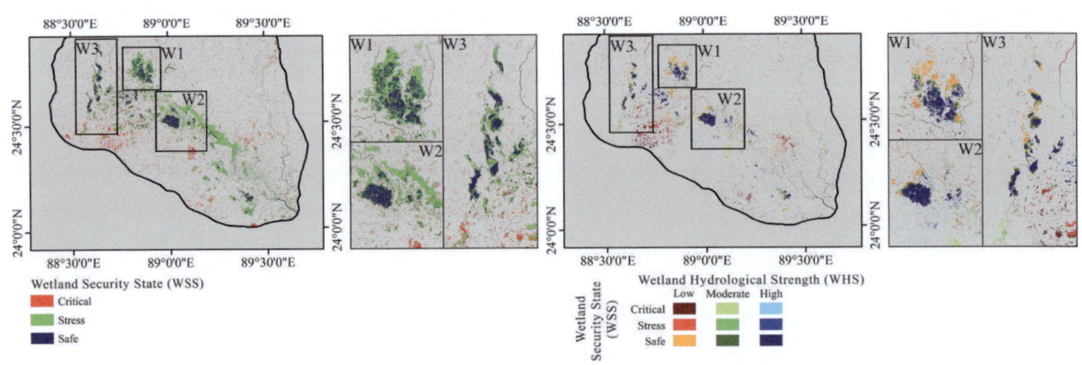

Fig. 7.2 Wetland hydrological security state (WSS) in relation to inundation water support and WSS and HS matrix which is showing HS of various WSS zones

curvature, spatial water budget, etc. are also very crucial (Frei et al. 2010). Here classification is based on only inundation parameter. So, scope is there to update the classification system, including other hydrological and associated parameters.

References

Al Amin MB, Haki H (2017) Floodplain simulation for Musi river using integrated 1D/2D hydrodynamic model. In: MATEC web of conferences, vol 101. EDP Sciences, p 05023

Bornette G, Amoros C (1996) Disturbance regimes and vegetation dynamics: role of floods in riverine wetlands. J Veg Sci 7(5):615–622

Casanova MT, Brock MA (2000) How do depth, duration and frequency of flooding influence the establishment of wetland plant communities? Plant Ecol 147:237–250

Chen L, Wu Y, Xu YJ, Zhang G (2021) Alteration of flood pulses by damming the Nenjiang River, China-Implication for the need to identify a hydrograph-based inundation threshold for protecting floodplain wetlands. Ecol Ind 124:107406

dos Santos Pinto RM, Weigelhofer G, Diaz-Pines E, Brito AG, Zechmeister-Boltenstern S, Hein T (2020) River-floodplain restoration and hydrological effects on GHG emissions: biogeochemical dynamics in the parafluvial zone. Sci Total Environ 715:136980

Frei S, Lischeid G, Fleckenstein JH (2010) Effects of micro-topography on surface–subsurface exchange and runoff generation in a virtual riparian wetland—a modeling study. Adv Water Resour 33(11):1388–1401

Freitag B, Bolton S, Westerlund F, Clark J (2012) Floodplain management: a new approach for a new era. Island Press

Liu Z, Zhao L, Xu T, Bu F, Liu X, Zhou D (2018) Quantification of potential flood inundation areas in the marsh wetland of Honghe National Natural Reserve, Northeast China. Ecohydrol Hydrobiol 18(4):355–364

Maurya PK, Malik DS, Yadav KK, Kumar A, Kumar S, Kamyab H (2019) Bioaccumulation and potential sources of heavy metal contamination in fish species in River Ganga basin: possible human health risks evaluation. Toxicol Rep 6:472–481

Middleton B (1999) Wetland restoration, flood pulsing, and disturbance dynamics. Wiley

Mitsch WJ (1992) Landscape design and the role of created, restored, and natural riparian wetlands in controlling nonpoint source pollution. Ecol Eng 1(1–2):27–47

Pal S, Let M (2024) Effect of hydrological modification on wetland morphology in reference to parts of the Bagri Region of Eastern India. In: Water resource management in climate change scenario: innovations in geospatial techniques and models. Springer, Cham, pp 221–234

Pal S, Talukdar S (2018) Drivers of vulnerability to wetlands in Punarbhaba river basin of India-Bangladesh. Ecol Ind 93:612–626

Quintela FM, Lima GP, Silveira ML, Costa PG, Bianchini A, Loebmann D, Martins SE (2019) High arsenic and low lead concentrations in fish and reptiles from Taim wetlands, a Ramsar site in southern Brazil. Sci Total Environ 660:1004–1014

Quirogaa VM, Kurea S, Udoa K, Manoa A (2016) Application of 2D numerical simulation for the analysis of the February 2014 Bolivian Amazonia flood: application of the new HEC-RAS version 5. Ribagua 3(1):25–33

Sarkar UK, Mishal P, Borah S, Karnatak G, Chandra G, Kumari S et al (2020) Status, potential, prospects, and issues of floodplain wetland fisheries in India: synthesis and review for sustainable management. Rev Fish Sci Aquaculture 29(1):1–32

Shen Y, Diplas P (2008) Application of two-and three-dimensional computational fluid dynamics models to complex ecological stream flows. J Hydrol 348(1–2):195–214

Talukdar S, Pal S (2020) Modeling flood plain wetland transformation in consequences of flow alteration in Punarbhaba river in India and Bangladesh. J Clean Prod 261:120767

Tockner K, Pennetzdorfer D, Reiner N, Schiemer F, Ward JV (1999) Hydrological connectivity, and the exchange of organic matter and nutrients in a dynamic river–floodplain system (Danube, Austria). Freshw Biol 41(3):521–535

Ward JV (1989) The four-dimensional nature of lotic ecosystems. J N Am Benthol Soc 8(1):2–8

Modeling Wetland Health

8

Abstract

Wetlands are ecologically sensitive systems that require robust assessment methodologies to quantify their health and degradation. This study tries to develop novel wetland health parameters by incorporating multi-sensor remote sensing datasets, integrating hydrological, thermal, and vegetation indices related indicators of wetland stability and degradation. Traditional machine learning (ML) algorithms (SVM, LR, RF, DT, GBM, XGB, LGBM) and advanced deep learning (DL) models (ANN, DNN, CNN, WDNN, SNN) were applied to assess wetland health before/pre- and after/post-dam construction periods. Model performance was assessed using different precision measures including recall, precision, F1-score, MCC, ROC-AUC, and Cohen's Kappa to ensure comprehensive assessment. The highest accuracy in pre-dam was observed in RF (0.848), XGB (0.850), and SNN (0.827), while in post-dam accuracy declined in all the models, with RF (0.828), XGB (0.830), and SNN (0.805) remaining top performers. The post-dam period exhibited increased wetland degradation, with the "very poor" class expanding significantly. Pre-dam estimated the highest at 219.06 km^2 (DT), while in post-dam prediction, it was found 169.50 km^2 (LGBM) for ML models (Table 1, Table 3) and 135.94 km^2 (WDNN) for DL models, indicating severe ecological decline. Concurrently, the "very good" wetland zones shrank considerably—from a pre-dam high of 539.43 km^2 (LGBM) to a post-dam maximum of only 128.05 km^2 (XGB) in ML predictions, and from 273.31 km^2 (WDNN) to just 94.82 km^2 (ANN) in DL models. These trends reflect a loss of wetland functionality and ecosystem integrity due to hydrological disruption. Explainable AI (XAI) analysis identified hydrological connectivity (HCS), water persistence (WPFI), and water stability (WEI) as primary indicators of resilience in the pre-dam scenario. In post-dam, dominant drivers shifted to anthropogenic stress (AESI), vegetation degradation (AVDI), and thermal stress (TWSI), indicating a transition from hydrological resilience to stress-induced deterioration. These findings underscore the adverse consequences of dam-induced hydrological alterations on wetland health and emphasize the urgent requirement for targeted protection strategies. Priorities should include restoration of hydrological connectivity, mitigation of vegetation loss, and control of thermal and anthropogenic pressures. The integration of ML, DL, and XAI provides a powerful framework for continuous wetland monitoring, supporting informed, data-driven policy for sustainable wetland management in human-impacted environments.

S. Pal et al., *Damming and Wetland Eco-hydrology*, GIScience and Geo-environmental Modelling,
https://doi.org/10.1007/978-3-032-11729-8_8

Keywords

Wetland health assessment · Machine and deep learning · Multi-sensor remote sensing · Explainable AI · Hydrological and ecological degradation · Pre- and post-dam wetland dynamics

8.1 Wetland Health Assessment Approaches

Traditional wetland assessment approaches have relied heavily on field-based surveys, which, although precise, are often resource-intensive, time-consuming, and spatially limited (Shuman and Ambrose 2003; Ghobadi et al. 2012). Consequently, remote sensing (RS) appeared as a vital wetland monitoring instrument, enabling large-scale, temporal, and cost-effective assessments (Ozesmi and Bauer 2002; Abdelmajeed et al. 2023).

Remote sensing methods have revolutionized wetland monitoring by facilitating the recognition and analysis of changes in wetland extent, vegetation health, hydrological patterns, and anthropogenic stresses (Jianping et al. 2007; Chen et al. 2014). Optical sensors, for instances those on MODIS, Landsat, and Sentinel-2, offer high-resolution RS imagery that allows for the classification and delineation of wetland types and conditions (Mahdavi et al. 2018; Lee and Yeh 2009). In addition, Sentinel-1-based Synthetic Aperture Radar (SAR) and RADARSAT have demonstrated remarkable efficiency in penetrating cloud cover and monitoring wetlands during different hydrological states (Kaplan and Avdan 2018; Orimoloye et al. 2019). The assimilation of multi-sensor RS data enhances the precision of wetland assessments, allowing researchers to develop robust models for predicting wetland health dynamics over time (Jafarzadeh et al. 2022; Shafi et al. 2023).

The appraisal of wetland health at a spatial scale requires a shift from conventional statistical models to advanced computational techniques that can handle high-dimensional and complex data structures (Demarquet et al. 2023). Traditional models, such as regression-based analyses, have been widely used but often not have the capability to capture intricate nonlinear relationships within wetland ecosystems (Shuman and Ambrose 2003; Zheng et al. 2022). The emergence of machine learning (ML) and deep learning (DL) algorithms provides an opportunity to improve wetland health modeling by leveraging large-scale datasets and extracting meaningful patterns with higher accuracy and efficiency (Yang et al. 2023; Gonzalez-Perez et al. 2022). ML approaches like gradient boosting decision trees (GBDT), support vector machines (SVM), and random forest (RF) were effectively applied for wetland categorization and health condition assessment (Ahmed et al. 2021; Martínez Prentice et al. 2021; Jamali et al. 2021; Demarquet et al. 2023). Furthermore, deep learning methods, including recurrent neural networks (RNNs) and convolutional neural networks (CNNs) have showed better performance in intricate image analysis tasks, enabling precise wetland mapping and degradation assessment (Jamali et al. 2021; Hosseiny et al. 2021; Günen 2022; Kaplan and Avdan 2018; Singh et al. 2024).

Despite the advancements in wetland monitoring using remote sensing and ML/DL models, numerous research gaps prevail. First of all, there is a requirement for developing advanced indicators of wetland health by assimilating multi-source RS datasets, including optical, SAR, and LiDAR, to enhance assessment accuracy (Das et al. 2022; Mondal et al. 2024). Secondly, existing ML models require further refinement to incorporate spatio-temporal dynamics and hydro-ecological parameters that better reflect wetland functionality (Aslam et al. 2024; Singh et al. 2024; Hasani et al. 2021; Yadav et al. 2024). Thirdly, paucity of explainability in ML/DL models prevails, raising concerns about their reliability in wetland management decisions. Explainable artificial intelligence (XAI) procedures, including SHapley Additive Explanations (SHAP), can perform a vital role in identifying the most influential parameters affecting wetland health (Gonzalez-Perez et al. 2022; Steinbach et al. 2023; Rawat et al. 2025).

Given these challenges, the present study intends to build up an advanced frame for wetland health assessment that integrates multi-source RS data with ML and DL approaches. The key objectives are to.

1. Develop novel wetland health parameters incorporating multi-sensor remote sensing datasets to construct advanced indicators of wetland condition and degradation.
2. Develop and apply traditional to advanced models for wetland health assessment.
3. Evaluate accuracy using multiple metrics for model performance.
4. Integrate XAI to determine key environmental and anthropogenic drivers influencing wetland health conditions.

By addressing these objectives, this chapter seeks to fill the gap between remote sensing applications and data-driven decision-making for sustainable wetland management. The paradigm shift from traditional statistical models to AI-driven methodologies offers promising avenues for improving wetland monitoring, ensuring proactive conservation strategies, and mitigating the effects of climate change and anthropogenic disturbances (Zheng et al. 2022; Mondal et al. 2024). Future investigation should spotlight on refining the models by including real-time sensor data, crowd sourced ecological observations, and participatory monitoring frameworks to enhance wetland resilience and ecosystem service sustainability (Yadav et al. 2024; Rawat et al. 2025).

8.2 Materials

The **data sources** for the study on wetland health assessment in the **Atreyee River Basin and its floodplain wetlands** were primarily derived from multi-sensor remote sensing datasets, leveraging both **MODIS and Landsat-based products** alongside ancillary datasets from **JRC, SRTM, and CIESIN GPWv4**. The integration of these datasets enables comprehensive hydrological and ecological assessments, capturing long-term trends and seasonal variability in wetland dynamics. The **MODIS Enhanced Vegetation Index (EVI) (MOD13A1)** and **Normalized Difference Water Index (NDWI) (MYD09GA_006_NDWI)** were utilized to assess **vegetation productivity and surface water extent**, respectively, reflecting wetland hydrological fluctuations. Furthermore, **MODIS Evapotranspiration (MOD16A2)** and **land surface temperature (MOD11A2) data** were incorporated to examine wetland hydro-thermal regimes and their spatio-temporal variations. The **JRC Global Surface Water dataset (GSW1_3)** was employed to quantify **long-term surface water occurrence**, which is critical for delineating wetland extents and identifying hydrological alterations. Additionally, **SRTM DEM (USGS/SRTMGL1_003)** provided topographic insights, aiding in the delineation of **hydrological catchments and floodplain dynamics**. To supplement these datasets, **Landsat-8 Collection 2 Level-2 (LC08/C02/T1_L2)** imagery was used for **water quality assessments,** leveraging **spectral indices for turbidity and chlorophyll estimation,** which serve as key indicators of wetland health. The **MODIS Land Cover (MCD12Q1)** dataset enabled an analysis of **land use transitions and wetland degradation patterns**. The study also integrated **population density estimates (CIESIN GPWv4)** to assess anthropogenic pressures, mapping **spatial correlations between human settlements and wetland vulnerability.** Data from **2002 to 2023** were processed using **Google Earth Engine (GEE),** wherein annual means were computed for each dataset to capture inter-annual variations. The methodological framework employed **image collection filtering, temporal aggregation, and spatial clipping** to extract relevant data for the **Atreyee River Basin**.

8.3 Methods

8.3.1 Methods for Computing Wetland Health Parameters

This study deploys a multi-sensor RS framework to measure hydrological and ecological degradation of floodplain wetlands amid Atreyee River

Basin. The methodology involves data acquisition, raster processing, and the computation of indices to quantify wetland health over pre-dam (2002–2012) and post-dam (2013–2024) phases.

8.3.1.1 Hydrological Degradation Indices

Water Persistence and Frequency Index (WPFI)

The Water Persistence and Frequency Index (WPFI) quantifies the consistency of surface water presence over multiple years, capturing long-term hydrological stability. It is derived from the NDWI time series, considering only positive values to reflect water presence:

$$\text{WPFI} = \frac{1}{T}\sum_{t=1}^{T}(\text{NDWI}_t > 0)\cdot \text{NDWI}_t, \quad (8.1)$$

where T = Number of images (of different years), NDWI_t = Normalized Difference Water Index at time t. The term ($\text{NDWI}_t > 0$) ensures only positive NDWI values contribute to the persistence measure.

A higher WPFI value indicates greater water permanence, while lower values suggest hydrological instability due to seasonal drying or anthropogenic modifications.

Water Degradation Intensity Index (WDII)

The WDII measures the rate of water degradation by computing the linear regression slope of NDWI over time, assessing the trend in wetland water loss:

$$\text{WDII} = |\text{Slope }(\text{NDWI}_t, t)|, \quad (8.2)$$

where the slope is derived from the linear regression equation:

$$\text{Slope} = \frac{T\sum t\text{NDWI}_t - \sum t \sum \text{NDWI}_t}{T\sum t_2 - \left(\sum t\right)^2}, \quad (8.3)$$

where t = Time step (year), NDWI_t = NDWI value at time t, T = Total number of years.

A higher WDII value signifies rapid water loss, indicating wetland deterioration due to hydrological modifications, climate change, or anthropogenic pressure.

Water Entropy Index (WEI)

The Water Entropy Index (WEI) quantifies the variability of NDWI values over time, capturing hydrological unpredictability:

$$\text{WEI} = -\sum_{i=1}^{n} p_i \log_2(p_i), \quad (8.4)$$

where p_i = Probability of NDWI values falling within histogram bin I, n = Number of histogram bins.

A higher WEI value indicates greater hydrological variability, which can lead to reduced wetland stability and increased flood risk.

Hydrological Connectivity Strength (HCS)

The HCS evaluates the spatial connectivity of water bodies, essential for sustaining wetland ecological functions. It is computed using an 8-neighborhood convolution filters applied to the water presence mask:

$$\text{HCS} = \frac{\sum \text{Convolve(Water Mask}, K)}{\sum \text{Water Mask}}, \quad (8.5)$$

where K = 3 × 3 kernel matrix for spatial connectivity, Water Mask = Binary mask of water presence.

Higher HCS values indicate strong hydrological connectivity, while lower values suggest wetland fragmentation due to hydrological modifications or land use changes.

8.3.1.2 Ecological Degradation Indices

Aquatic Vegetation Degradation Index (AVDI)

The AVDI integrates vegetation health, algal growth, and water turbidity to assess aquatic ecosystem stress:

$$\begin{aligned}\text{AVDI} =& 0.4|\text{Slope}(\text{EVI}_t, t)| + 0.4\max(\text{Chlorophyll}) \\ &+ 0.2\sigma(\text{Turbidity}), \end{aligned} \quad (8.6)$$

where $|\text{Slope}(\text{EVI}_t, t)|=$ Trend in enhanced vegetation index over time, $\max(\text{Chlorophyll}) =$ Maximum chlorophyll concentration, $\sigma(\text{Turbidity}) =$ Standard deviation of turbidity values.

Higher AVDI values indicate vegetation stress, often due to increased sedimentation, eutrophication, or water pollution.

Thermal-Water Stress Index (TWSI)

The TWSI assesses combined effects of water stress, temperature variations, and elevation influence:

$$\begin{aligned}\text{TWSI} =&0.4|\text{Slope}(\text{NDWI}_t, t)\mid +0.4\sigma(\text{LST})\\ &+0.2\frac{\text{Elevation} - \min(\text{Elevation})}{\max(\text{Elevation}) - \min(\text{Elevation})},\end{aligned} \tag{8.7}$$

where $\sigma(\text{LST}) =$ Standard deviation of land surface temperature, Elevation = Normalized elevation values.

Higher TWSI values indicate increased temperature-induced water stress, leading to wetland ecosystem vulnerability.

Water Pollution Risk Index (WPRI)

The WPRI integrates water quality degradation factors:

$$\begin{aligned}\text{WPRI} =&0.5\max(\text{Chlorophyll})\\ &+0.3\mid \text{Slope}(\text{Turbidity}_t, t)\mid\\ &+0.2(1 - \text{WPFI}),\end{aligned} \tag{8.8}$$

where $\max(\text{Chlorophyll}) =$ Peak chlorophyll-a concentration (indicator of algal blooms), $\text{Slope}(\text{Turbidity}_t, t) =$ Trend in turbidity over time, WPFI = Water Persistence and Frequency Index.

Higher WPRI values signify greater pollution risks, primarily from nutrient enrichment and sediment influx.

Sediment Transport Index (STI)

The STI quantifies sediment dynamics using turbidity intensity and elevation gradient:

$$\text{STI} = 0.6\max(\text{Turbidity}) + 0.4\frac{\nabla\text{Elevation}}{\max(\nabla\text{Elevation})}, \tag{8.9}$$

where $\max(\text{Turbidity}) =$ Maximum turbidity levels, $\nabla\text{Elevation} =$ Gradient of elevation.

Higher STI values indicate active sediment transport, often associated with flooding, erosion, or human-induced disturbances.

Aquatic Ecosystem Stability Index (AESI)

The AESI integrates NDWI, LST, and chlorophyll stability to assess overall wetland ecosystem health:

$$\begin{aligned}\text{AESI} =&0.4(1 - \sigma(\text{NDWI})) + 0.3(1 - \sigma(\text{LST}))\\ &+0.3(1 - \sigma(\text{Chlorophyll})),\end{aligned} \tag{8.10}$$

where $\sigma(\text{NDWI}) =$ Standard deviation (SD) of NDWI, $\sigma(\text{LST}) =$ Standard deviation (SD) of LST, $\sigma(\text{Chlorophyll}) =$ Standard deviation of chlorophyll-a.

Higher AESI values indicate greater ecosystem stability, while lower values reflect high fluctuations and environmental stress.

8.3.1.3 Rationale for Selecting Parameters

The selection of hydrological and ecological parameters was based on their relevance in capturing wetland degradation trends in the Atreyee Basin in pre- and post-Mohanpur dam building. The dam, completed in 2012, significantly altered hydrological connectivity, water retention, and ecosystem stability, necessitating a comparative assessment across pre-dam (before dam) (2002–2012) and post-dam(after dam) (2013–2024) phases.

Hydrological indices were selected to quantify water stability, degradation trends, and connectivity, given the altered discharge regime and floodplain contraction in post-dam. WPFI was chosen to evaluate long-term water retention, as the dam disrupted natural seasonal inundation cycles, reducing monsoonal flood intensity. A declining WPFI post-dam would indicate loss of perennial wetlands, leading to hydrological fragmentation. WDII was selected to track surface water depletion trends, as the regulated flow regime and sediment retention behind the dam likely contributed to accelerate drying of floodplain wetlands. WEI was included to assess fluctuations in water stability, as increased hydrological unpredictability in post-dam could be attributed to reduce seasonal flooding and localized groundwater dependence. HCS was incorporated to quantify water network continuity, given that reduced discharge in post-dam fragmented wetland hydrological connections, potentially impacting biodiversity and aquatic habitat availability.

Ecological indices were chosen to evaluate vegetation health, pollution risk, and ecosystem stability, recognizing that post-dam hydrological stressors influenced vegetation productivity and water quality. AVDI was selected to assess aquatic vegetation degradation, as declining water availability, increased sedimentation, and nutrient loading post-dam likely affected wetland primary productivity. TWSI was included to capture temperature-induced wetland stress, given that water regulation altered microclimatic conditions, leading to higher LST variability post-dam. WPRI was introduced to examine water pollution risks, considering that reduced discharge and increased agricultural runoff post-dam could have intensified eutrophication and sediment deposition in wetlands. STI was selected to quantify sediment transport changes, as reduced river velocity and altered flood patterns likely modified sedimentation dynamics and erosion-prone areas. Finally, AESI was included to provide a comprehensive measure of wetland resilience, capturing multidimensional stressors across water, temperature, and biological stability.

The pre-dam period was characterized by relatively unregulated monsoonal flooding, natural wetland connectivity, and stable hydrological regimes, making WPFI, WEI, and HCS critical for assessing baseline wetland health. In post-dam, hydrological restrictions, reduced flood pulse, and increased anthropogenic stressors altered wetland functions, necessitating a strong focus on degradation trends through WDII, TWSI, WPRI, and AVDI. This parameter selection ensures a comprehensive, quantitative assessment of wetland transformations, aiding in evidence-based conservation and adaptive management strategies for the Atreyee River Basin.

8.3.2 Application of Traditional to Advanced Computation Techniques

The assessment of wetland health conditions using remote sensing-based models has undergone a significant transition from traditional statistical techniques to advanced artificial intelligence (AI), incorporating machine and deep learning (DL). The study applied traditional ML models such as logistic regression (LR), random forest (RF), decision tree (DT), support vector machine (SVM), gradient boosting (GB), extreme gradient boosting (XGB), and light GBM (LGBM) to classify wetland degradation, followed by the deployment of advanced DL architectures like artificial neural network (ANN), convolutional neural network (CNN), deep neural network (DNN), wide and deep neural network (WDNN), and self-normalizing network (SNN) for enhanced pattern recognition. The transition from traditional ML to advanced DL approaches ensures better generalization, improved spatial prediction, and enhanced interpretability of wetland degradation trends.

8.3.2.1 Inventory Creation for Machine Learning and Deep Learning Training

A structured dataset was created integrating spectral remote sensing indices, field surveys, and wetland fragmentation analyses. The study length was categorized into pre-dam (2002–2012) and post-dam (2013–2024) phases, with presence (healthy wetlands) and absence (degraded wetlands) being classified based on ground-truth observations and spatial overlays of hydrological indices. The inventory was prepared by integrating multi-sensor datasets (MODIS, Landsat-8, and JRC Global Surface Water) and deriving hydrological (NDWI, WPFI, WDII), vegetation (EVI, AVDI), water quality (TWSI, WPRI), and connectivity (HCS, STI) indices. Raster values were extracted for sample locations using Geopandas and Rasterio, ensuring that the dataset maintained spatial and temporal consistency. The processed raster values were merged into structured CSV files, facilitating their application in ML and DL training workflows.

8.3.2.2 Traditional Machine Learning Models for Wetland Degradation Prediction

Logistic Regression (LR)

Logistic regression (LR), a binary classification approach, is widely employed for environmental hazard prediction, including wetland degradation assessment (Sahani and Ghosh 2021). The model presumes a linear association between analyst variables and the probability of degradation, building it helpful for interpretable classification tasks (Niaz et al. 2021). However, its inability to model complex, nonlinear interactions limits its predictive capabilities in spatially heterogeneous wetland environments (Gu et al. 2022).

Support Vector Machine (SVM)

SVM, a nonlinear classifier, separates degraded and stable wetland areas using hyperplane-based classification (Wang 2022). It is particularly effective in identifying spatially complex degradation patterns when combined with kernel functions (Sahu and Pandey 2023). However, SVM computationally scales poorly with large datasets, making it less efficient for high-resolution wetland classification tasks (Ekmekcioğlu and Koc 2022).

Decision Tree (DT) and Random Forest (RF)

Decision trees (DT) generate rule-based classification models, while random forest (RF) enhances accuracy by combining multiple bootstrapped decision trees (Pham et al. 2022). RF is particularly robust to outliers and nonlinearity, making it suitable for predicting wetland degradation (Talebi et al. 2022). However, over fitting in decision trees can lead to classification errors, requiring regularization techniques for performance improvement (Nafouanti et al. 2021).

Gradient Boosting (GB), Extreme Gradient Boosting (XGB), and LightGBM (LGBM)

Bharti et al. (2021) rightly pointed out that gradient boosting (GB) is an ensemble method that optimizes weak learners sequentially, significantly improving predictive accuracy for wetland degradation mapping. Extreme gradient boosting (XGB) refines the GB approach using parallelized boosting and advanced loss functions, making it an efficient choice for wetland classification (Sahin 2022). LightGBM (LGBM) further enhances XGB's performance by reducing computational costs and handling large datasets efficiently (Kavzoglu and Teke 2022). These models have demonstrated high classification accuracy in predicting wetland stressors, sedimentation trends, and vegetation degradation (Li et al. 2024; Saber et al. 2022).

8.3.2.3 Advanced Deep Learning Models for Wetland Degradation Prediction

Artificial Neural Network (ANN)

Artificial neural networks (ANNs) leverage multilayers of interrelated neurons to model nonlinear interactions among wetland indices (Darabi et al. 2021). Unlike ML models, ANN automatically extracts hierarchical features, enabling better classification of hydrological and ecological degradation (Guha et al. 2022). The major constraint of ANN is its susceptibility to over fitting, requiring regularization techniques for improved performance.

Deep Neural Network (DNN)

Deep neural networks (DNNs) expand ANNs by including multi-hidden layers, permitting them to extract intricate spatial patterns in wetland degradation (Yao et al. 2022). DNNs have been widely used in climate-induced disaster prediction and hydrological impact modeling (Yousefzadeh et al. 2021). The study implemented dropout layers and batch normalization to improve model generality, ensuring better classification of post-dam wetland conditions.

Convolutional Neural Network (CNN)

Convolutional neural networks (CNNs) gained popularity in spatio-temporal environmental assessments, especially in wetland and hydrological modeling (Van Le et al. 2021). CNNs leverage spatial feature extraction techniques using convolutional filters, allowing the model to learn hydrological connectivity patterns (Wei et al. 2021). This study used 1D convolutional layers to detect temporal variations in wetland indices, improving classification performance compared to traditional ML approaches.

Wide and Deep Neural Network (WDNN)

WDNN combines deep learning-based feature extraction with wide regression-based interpretability, making it an effective model for wetland degradation assessment (Zhao et al. 2025). WDNN enhances spatial dependency modeling and improves classification precision, especially for fragmented wetlands (Liu et al. 2022). By integrating batch normalization and dropout layers, WDNN effectively mitigated bias in post-dam assessments.

Self-normalizing Neural Network (SNN)

SNNs utilize Scaled Exponential Linear Unit (SELU) activation functions to enable self-normalizing transformations, making them suitable for long-term hydrological impact modeling (Alatrany et al. 2023). SNN was implemented in the study to predict degradation risk, considering non-stationary hydrological fluctuations and climatic variability (Madasu and Rao 2019).

8.3.2.4 Model Optimality Assessment

Ensuring optimal model performance for wetland health assessment requires a systematic evaluation of predictive stability, generalization ability, and computational efficiency. The study used a multi-metric approach to assess the reliability of traditional ML and advanced DL models using a combination of accuracy-driven and probabilistic evaluation techniques. Traditional ML models, including LR, SVM, DT, RF, GB, XGB, and LightGBM, were subjected to hyperparameter tuning using grid search and cross-validation to enhance classification robustness (Sahani and Ghosh 2021; Pham et al. 2022). The hyperparameter space included regularization parameters, tree depth, kernel types, and learning rate schedules, ensuring that each model was optimized for non-stationary hydrological conditions observed in pre- and post-dam datasets (Niaz et al. 2021). ML model performance was first assessed using hold-out validation on test datasets, followed by probabilistic output validation using area under curve (AUC) under receiver operating characteristic (RoC), providing an estimate of model confidence under uncertain wetland conditions (Gu et al. 2022). To further validate classification stability, ensemble averaging was implemented for tree-based models, allowing RF, GB, and XGB to demonstrate improved generalization and lower over fitting risk (Talebi et al. 2022).

For deep learning models, including ANN, DNN, CNN, WDNN, and SNN, model training involved an adaptive learning rate schedule, employing the Adam optimizer to dynamically adjust weight updates (Yao et al. 2022). DL model architectures were assessed for computational efficiency and feature abstraction depth, with CNN leveraging localized feature extraction from raster datasets, while DNN and WDNN focused on hierarchical feature learning for wetland degradation classification (WEI et al. 2021). To improve model resilience against over fitting, techniques such as dropout regularization, batch normalization, and early stopping were introduced (Darabi et al. 2021). These measures ensured stable gradient propagation across multiple hidden layers, reducing the risk of exploding or diminishing gradients that are common concerns in deep-layer architectures (Zhao et al. 2025). For each DL model, learning curves were analyzed by plotting epoch-wise accuracy trends, assessing how models improved with progressive training iterations while monitoring for convergence stability (Liu et al. 2022). The combination of hyperparameter tuning, adaptive learning mechanisms, and ensemble strategies ensured that both ML and DL models reached optimal predictive configurations suited for spatio-temporal wetland assessment.

8.3.2.5 Model Performance Evaluation

Once the models were optimally trained, performance evaluation was conducted using a multi-indicator statistical framework, ensuring a comprehensive assessment of predictive reliability, classification robustness, and decision consistency. The evaluation metrics included accuracy components including recall, precision, F1-score, ROC, Matthews correlation coefficient (MCC), balanced accuracy, Cohen's Kappa, Jaccard score, and log loss. Accuracy provided a basic correctness measure, while precision-recall trade-offs were crucial for identifying model sensitivity to class imbalance in wetland degradation classification (Talebi et al. 2022). Harmonic mean of recall and precision is F1-score which quantified the exchange between false negatives and false positives, ensuring that models provided reliable wetland classification under uncertain hydrological conditions (Pham et al. 2022). ROC-AUC analysis assessed model confidence by computing true positive and false positive trade-offs, with higher ROC-AUC values indicating a better differentiation between degraded and stable wetlands (Kavzoglu and Teke 2022). To evaluate spatial consistency in classification performance, heatmap visualizations of model-wise performance metrics were generated, allowing comparison between pre-and post-dam phases. Visualization provided insights into which models generalized well across different hydrological conditions and which exhibited classification instability due to noise or overfitting (Sahin 2022). Among traditional ML models, RF, XGB, and LGBM demonstrated superior performance due to their capacity to lever high-dimensional hydrological feature sets and capture nonlinear degradation patterns (Saber et al. 2022). Conversely, LR and DT exhibited relatively lower performance, reflecting their limited ability to model spatial dependencies in wetland degradation indicators (Nafouanti et al. 2021). In the DL domain, CNN and WDNN consistently outperformed ANN and SNN models, demonstrating their superior capability to extract hierarchical spatial relationships from remote sensing-derived wetland indices (WEI et al. 2021). The CNN model, leveraging convolutional filters for feature abstraction, achieved the maximum accuracy and F1-score, building it the most suitable model for large-scale wetland degradation mapping (Van Le et al. 2021).

An additional layer of performance evaluation involved analyzing error distribution using confusion matrices, highlighting model-specific false classification tendencies. The post-dam dataset exhibited higher classification complexity, requiring models with enhanced feature abstraction capacity (Zhao et al. 2025). Therefore, ensemble-based ML methods (XGB, LGBM, RF) and deep learning architectures (CNN, WDNN) demonstrated higher predictive reliability, effectively capturing spatial degradation trends. The final model selection was based on a weighted multi-metric ranking, ensuring that the highest-performing models in accuracy, precision-recall balance, and spatial generalization were considered optimal for wetland vulnerability assessments in the Atreyee River Basin.

8.3.3 Methods for Going Beyond Health Condition Modeling Using ML and DL

The integration of Explainable AI (XAI) frameworks in wetland health condition modeling enables a transparent, interpretable, and accountable decision-making process by going beyond conventional black-box ML and DL models. This study leverages SHapley Additive exPlanations (SHAP), confidence variability assessments, permutation importance, model agreement matrices, and robustness analysis to evaluate how individual predictors influence wetland degradation classifications. SHAP values provide a game-theoretic approach to quantifying feature importance, ensuring that the models' decisions align with ecological and hydrological knowledge. In this implementation, SHAP computes local and global feature attributions by measuring the effect of every feature on the projected probability of wetland health (Van Le et al. 2021; Wei et al.

2021). This method is critical for understanding nonlinear dependencies between remote sensing indicators and wetland conditions. Additionally, permutation importance, which randomly shuffles individual features and evaluates their influence on model performance, ensures that hydrological indices such as WPFI, WDII, and TWSI are appropriately weighted in predictive models (Yao et al. 2022). To extend interpretability to spatial modeling contexts, feature importance heat maps were generated for pre- and post-dam phases, visualizing how the significance of hydrological, vegetation, and connectivity indices evolved over time.

To further enhance interpretability, model agreement matrices were computed to measure the consistency of predictions across different ML and DL models, revealing areas where ensemble-based models (RF, GB, XGB, and LGBM) exhibited higher predictive agreement compared to ANN, DNN, and CNN (Zhao et al. 2025). This agreement assessment is crucial for validating the stability of AI-based predictions under varying ecological conditions (Liu et al. 2022). The confidence variability of models was analyzed using the standard deviation of prediction probabilities, indicating how certain or uncertain a model is when classifying wetland health conditions. DL models demonstrated higher confidence levels than traditional ML models due to their capacity to pull out hierarchical spatial attributes from multi-temporal data. However, higher confidence does not always equate to higher robustness; therefore, model robustness was assessed by introducing Gaussian noise perturbations to the test datasets and analyzing the impact on predictive stability (Darabi et al. 2021). Robustness scores provided insights into how ML and DL models adapted to uncertainties in wetland monitoring, with CNN and WDNN demonstrating superior resilience to perturbations, making them more reliable for long-term wetland sustainability assessments. The combined use of SHAP, permutation importance, model agreement matrices, confidence variability, and robustness metrics establishes an advanced XAI framework, ensuring that predictions are not only accurate but also interpretable, trustworthy, and ecologically meaningful.

8.4 Results

8.4.1 Spatial Assessment of Wetland Health Parameters

The spatio-temporal distribution of wetland health parameters in the pre-dam condition reveals significant variations in hydrological and ecological integrity over the study unit (Fig. 8.1). The Water Persistence and Frequency Index (WPFI) (Fig. 8.1a) demonstrates high values (~ 0.187) in the central and western sections, indicating regions with sustained water presence, which is crucial for wetland stability. However, lower WPFI values dominate the eastern and peripheral zones, signifying areas prone to seasonal desiccation, making them more vulnerable to hydrological stress and habitat loss. Similarly, the Water Degradation Intensity Index (WDII) (Fig. 8.1b) shows peak values (~ 0.09) concentrated in the south western and central floodplain, suggesting regions where gradual hydrological depletion has occurred over time. This indicates that while some areas exhibit stable wetland conditions, others are undergoing progressive water degradation, leading to fragmentation and drying of wetland habitats. The Water Entropy Index (WEI) (Fig. 8.1c) highlights variations in water stability, with high values (~ 3.27) along the central drainage network and lower reaches, suggesting fluctuating hydrological patterns. These regions are characterized by dynamic water retention, making them more sensitive to external environmental perturbations. Conversely, the Hydrological Connectivity Strength (HCS) (Fig. 8.1d) exhibits high connectivity values (~ 0.6) across the western and central wetlands, indicating a well-connected hydrological network that facilitates efficient water exchange. In contrast, fragmented sections in the eastern domain exhibit lower connectivity, suggesting reduced water flow, which may lead to stagnation and localized wetland degradation. The Aquatic Vegetation Degradation Index (AVDI) (Fig. 8.1e) indicates severe vegetation stress, with high degradation values (~ 87.25) across northern and central wetland patches, implying extensive loss

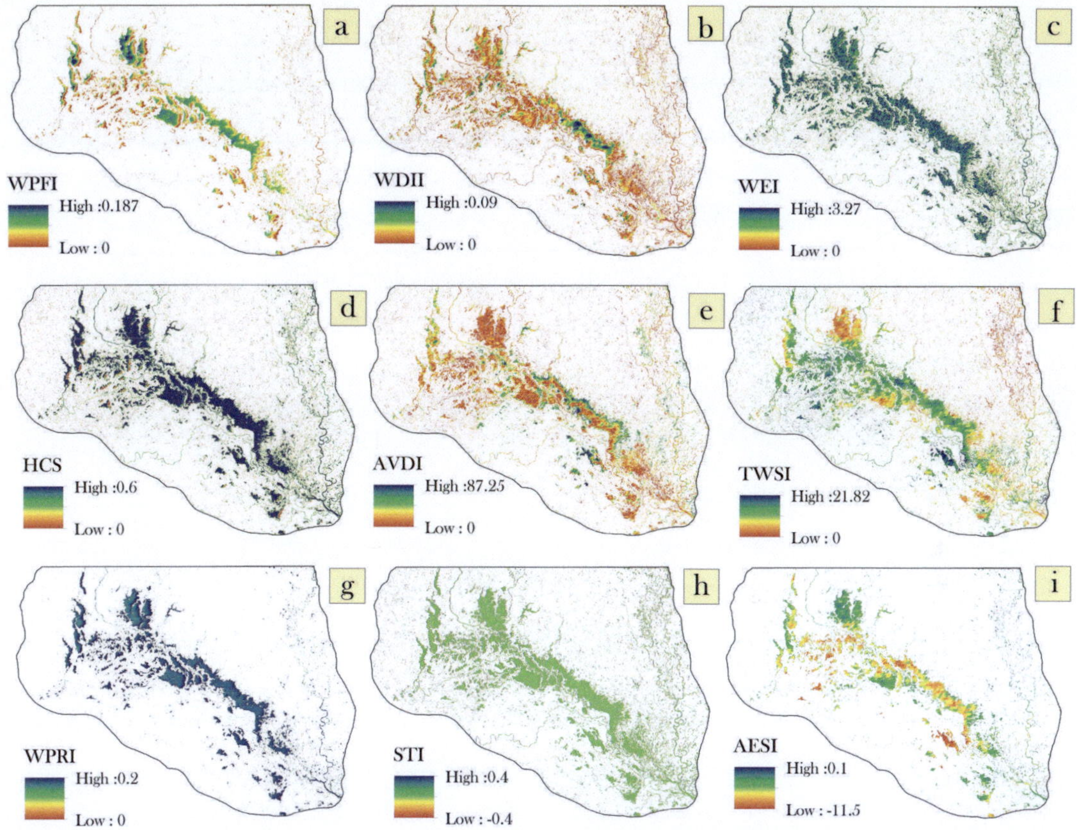

Fig. 8.1 Spatial distribution of wetland health parameters in pre-dam condition **a** WPFI, **b** WDII, **c** WEI, **d** HCS, **e** AVDI, **f** TWSI, **g** WPRI, **h** STI, **i** AESI

of aquatic vegetation cover. This pattern aligns with regions experiencing reduced hydrological support, which significantly impacts ecosystem stability. Similarly, the Thermal-Water Stress Index (TWSI) (Fig. 8.1f) exhibits peak values (~ 21.82) in the central and south western floodplain, indicating areas subjected to thermal stress due to water scarcity and increased evapotranspiration. This suggests that these wetlands are vulnerable to temperature-induced hydrological shifts, which may further exacerbate drying trends. The Water Pollution Risk Index (WPRI) (Fig. 8.1g) shows higher contamination risk (~ 0.2) in the south western and western wetland sections, where hydrological retention is prolonged, leading to accumulation of pollutants. This suggests a potential decline in water quality, which could diminish wetland productivity and increase susceptibility to eutrophication. The Sediment Transport Index (STI) (Fig. 8.1h) highlights moderate sediment transport (~ 0.4) in upstream regions, where increased flow energy supports sediment mobility, while downstream wetlands exhibit low STI values, indicating sediment deposition zones that contribute to habitat formation. Lastly, the Aquatic Ecosystem Stability Index (AESI) (Fig. 8.1i) reveals highly unstable zones (– 11.5) in the eastern and south western sections, suggesting degraded ecological conditions due to hydrological fragmentation and anthropogenic alterations.

The spatial assessment of wetland health parameters in the post-dam condition (Fig. 8.2) reveals significant hydrological and ecological degradation compared to the pre-dam scenario (Fig. 8.1). The WPFI (Fig. 8.2a) shows a marginal increase in water persistence (~ 0.197) in the central wetland zones, but the peripheral regions

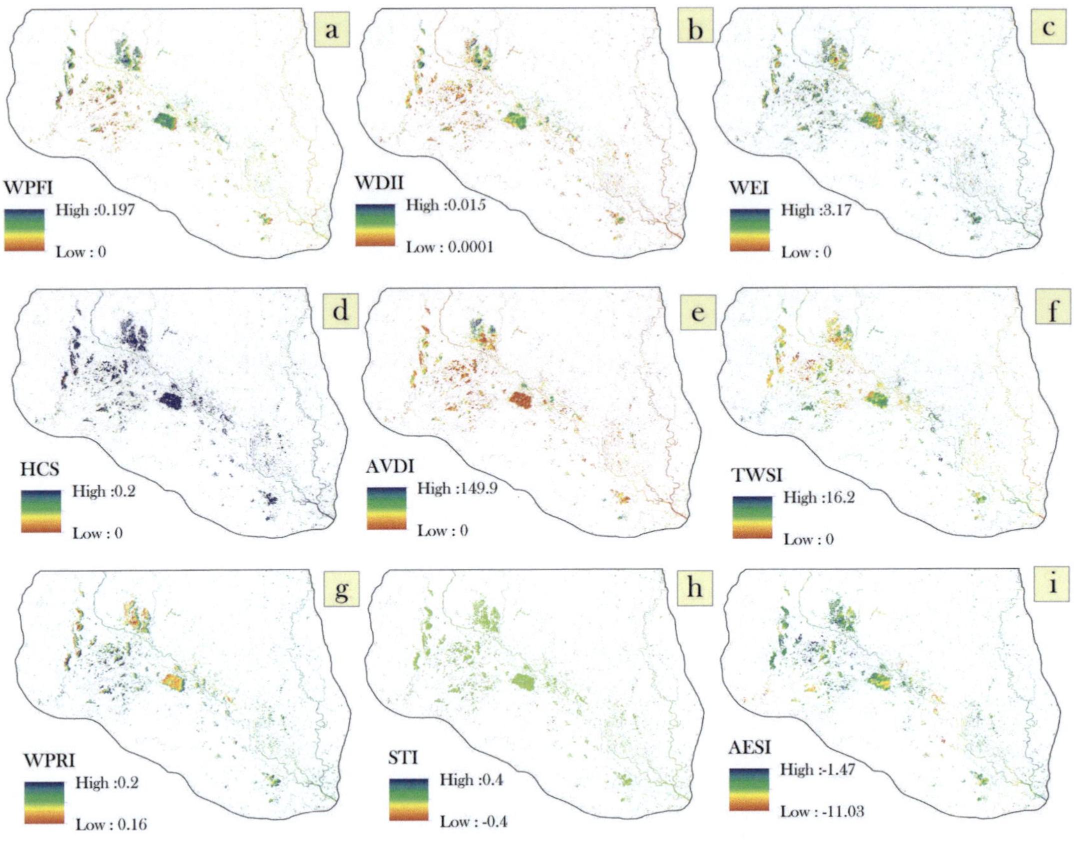

Fig. 8.2 Spatial distribution of wetland health parameters in post-dam condition **a** WPFI, **b** WDII, **c** WEI, **d** HCS, **e** AVDI, **f** TWSI, **g** WPRI, **h** STI, **i** AESI

exhibit further reduction in water retention, indicating intensified seasonal drying. The WDII (Fig. 8.2b) has decreased (~ 0.015), suggesting that while some areas still face water depletion, the rate of degradation has reduced slightly due to residual hydrological stabilization. However, the WEI (Fig. 8.2c) remains nearly unchanged (~ 3.17), indicating that fluctuations in water availability continue to affect wetland stability, with localized patches of severe variability in the eastern floodplain. A critical decline is observed in HCS (Fig. 8.2d), which has dropped to 0.2, indicating a substantial disruption in hydrological connectivity across the wetland system. This suggests that the dam's influence has intensified water fragmentation, reducing natural water exchange mechanisms. The AVDI (Fig. 8.2e) exhibits a drastic increase (~ 149.9), indicating severe aquatic vegetation degradation, particularly in the southern and central wetland patches. This is notably worse than pre-dam conditions, as hydrological alterations have significantly impacted vegetative resilience and wetland productivity. Similarly, TWSI (Fig. 8.2f) remains high (~ 16.2), confirming that thermal-induced stress due to evapotranspiration and reduced water buffering capacity has intensified in post-dam conditions, particularly in northeastern zones. The WPRI (Fig. 8.2g) maintains a high value (~ 0.2) with increased pollution risks in downstream regions, likely due to longer retention of contaminants and reduced natural flushing mechanisms post-dam. The STI (Fig. 8.2h) remains unchanged (~ 0.4), but its spatial distribution has shifted, with increased sediment deposition observed in upstream zones, indicating further modification of sediment transport processes. The most alarming decline is observed in AESI (Fig. 8.2i), which has dropped to − 11.03, suggesting worsened ecological instability. This reinforces the notion that post-dam wetland conditions have severely deteriorated, with reduced resilience to hydrological perturbations and increased ecosystem fragility.

A direct comparison between pre- and post-dam conditions highlights a worsening trend across all critical wetland health indicators. HCS has significantly declined, restricting natural water movement and leading to localized stagnation. Vegetation degradation (AVDI) has intensified, with post-dam conditions showing a nearly twofold increase in degradation intensity. Thermal stress (TWSI) and water pollution risk (WPRI) remain high, suggesting that the dam has altered wetland hydrodynamics, exacerbating temperature-related stress and pollutant accumulation. The overall ecological stability (AESI) has drastically declined, indicating that the post-dam wetland system is highly vulnerable to further degradation. These findings underscore the negative impact of hydrological modifications on wetland sustainability, necessitating urgent mitigation strategies to restore ecosystem functions.

8.4.2 Assessment of Wetland Health Conditions Using Traditional to Advanced Computation

The assessment of wetland health conditions necessitates the amalgamation of traditional and advanced computational procedures to augment predictive accuracy and interpretability. The evaluation of model optimality is crucial to ensuring robust wetland vulnerability assessments, particularly in the context of pre- and post-dam conditions. By comparing ML and DL model performance, insights into the reliability, precision, and generalizability of computational models can be obtained, facilitating superior decision-framing for wetland preservation and management.

8.4.3 Evaluation of Model's Optimality

Model's Optimality Assessment for Pre-dam Conditions

The performance evaluation of ML models (Fig. 8.3) for pre-dam conditions indicates a high degree of accuracy across multiple classifiers, with RF, GB, and XGB achieving the highest accuracy (~ 0.82), followed closely by LGBM. In contrast, LR and SVM exhibit slightly lower accuracy (~ 0.79), likely due to their linear nature, which may not effectively capture the complex nonlinear dependencies inherent in wetland health dynamics. The MSE values further support this observation, as tree-based ensemble models (RF, GB, XGB, and LGBM) demonstrate lower error rates compared to LR and SVM. The superior performance of RF and GB could be attributed to their capability to grip high-dimensional data, account for interactions between variables, and mitigate over fitting through bagging and boosting mechanisms, respectively (Fig. 8.3a, b). Similarly, the DL models (Fig. 8.4) exhibit comparable trends, with SNN attaining the highest accuracy (~ 0.86), followed by CNN (~ 0.84), while ANN and WDNN maintain slightly lower but competitive accuracy values (~ 0.81). The advanced architectures of DNN, WDNN, and SNN allow them to capture intricate spatial–temporal dependencies in wetland health indicators, leading to improved classification outcomes. The observed MSE values corroborate this trend, with CNN and SNN achieving lower error rates than ANN and WDNN (Fig. 8.4a, b). These outcomes underscore the efficiency of DL models in modeling wetland health conditions, where complex relationships between hydrological and ecological parameters necessitate feature hierarchies and deep feature extraction. The learning curves of DL models (Fig. 8.5) provide additional insights into model optimality by visualizing convergence trends and stability over multiple epochs. The SNN model exhibits the

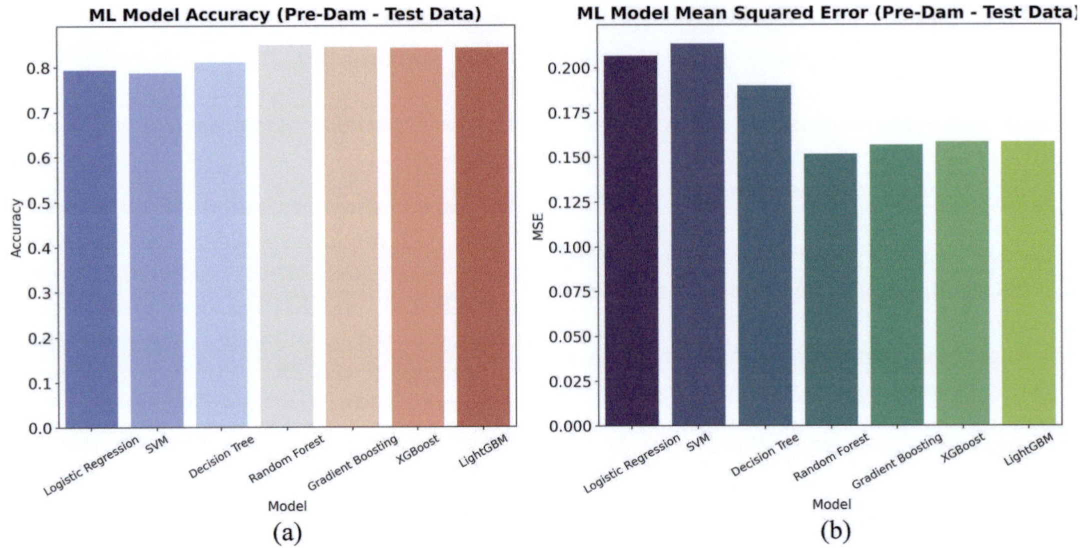

Fig. 8.3 Performance of ML models in assessing wetland health conditions during pre-dam conditions with **a** accuracy, **b** MSE values

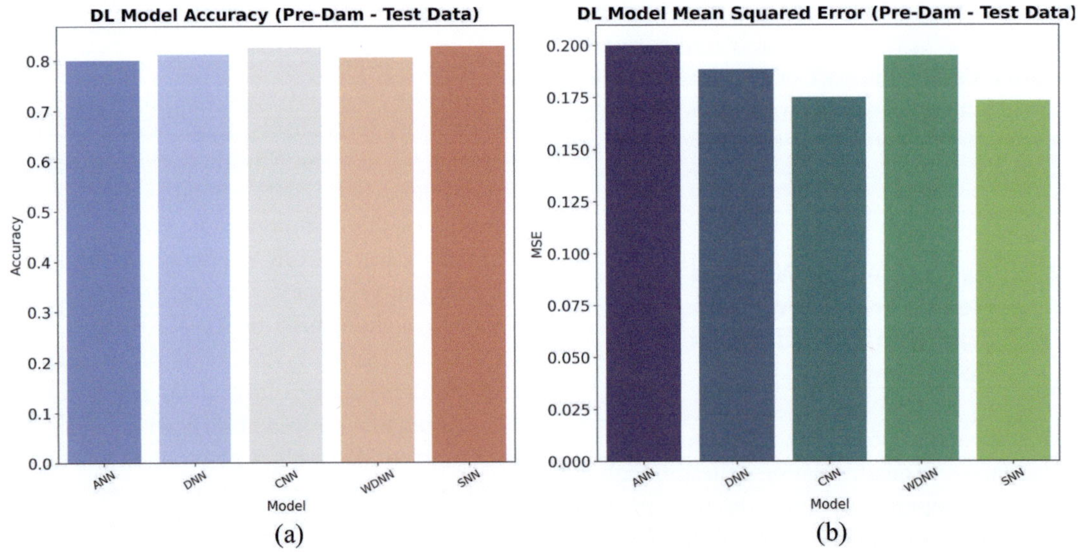

Fig. 8.4 Performance of DL models in assessing wetland health conditions during pre-dam conditions with **a** accuracy, **b** MSE values

most stable learning curve, with minimal oscillations in validation accuracy, suggesting enhanced generalizability. Conversely, ANN and WDNN show moderate fluctuations, indicating sensitivity to hyperparameter tuning and potential issues of over fitting in specific instances. The CNN model demonstrates steady improvement in accuracy over epochs, reinforcing its suitability for spatial pattern recognition in wetland health assessment. These observations imply that DL models, particularly CNN and SNN, effectively capture latent patterns in multi-sensor remote sensing datasets, thereby outperforming traditional ML approaches in predictive accuracy and reliability.

The relative evaluation of ML as well as DL models highlights the progressive swing toward

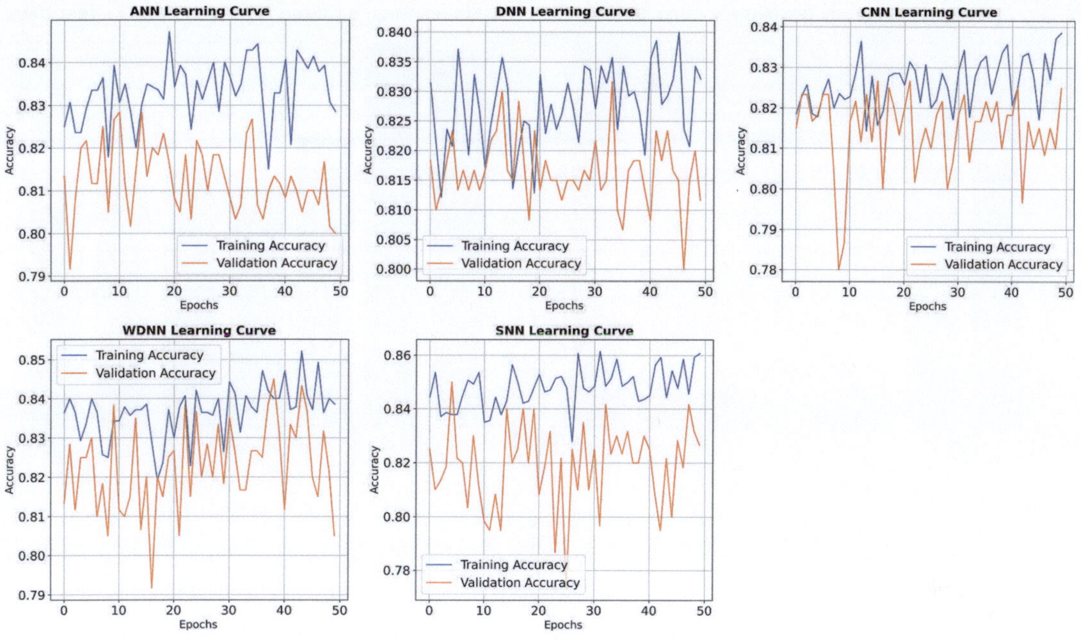

Fig. 8.5 Learning curves of DL models in pre-dam conditions

further sophisticated computational approaches in wetland health assessment. The lower MSE values and higher accuracy scores obtained by tree-based ML models (RF, GB, XGB) and DL architectures (CNN, SNN) indicate their superior predictive performance in detecting wetland degradation patterns. This finding has significant implications for large-scale environmental monitoring, as advanced DL models can leverage spatial–temporal feature learning to improve classification robustness, even in highly dynamic wetland ecosystems. Furthermore, the learning curve analysis suggests that while ANN and WDNN demonstrate moderate instability in validation accuracy, CNN and SNN maintain consistent performance, emphasizing the importance of model architecture in optimizing wetland health prediction. The improved model stability observed in CNN and SNN highlights their potential applicability in long-term wetland vulnerability assessments, enabling policymakers and conservationists to make informed decisions based on reliable computational predictions.

Model's Optimality Assessment For Pre-dam Conditions

The post-dam wetland condition assessment necessitates a comparative evaluation of ML and DL model performance to determine the optimal computational framework for wetland health monitoring. The post-dam scenario introduces increased variability and complexity in wetland dynamics, making it imperative to assess model stability, accuracy, and generalization ability. The evaluation of ML and DL models is critical for identifying the most robust approach for predicting hydrological and ecological degradation in the altered wetland system. The ML models demonstrate varying levels of predictive capability in the post-dam period (Fig. 8.6). The accuracy of models such as RF, GB, and XGB remains consistently high, exceeding 80%, while LR and SVM exhibit relatively lower performance, with accuracy values closer to 75%. The MSE analysis further reveals that LR and SVM suffer from higher error values, indicating suboptimal

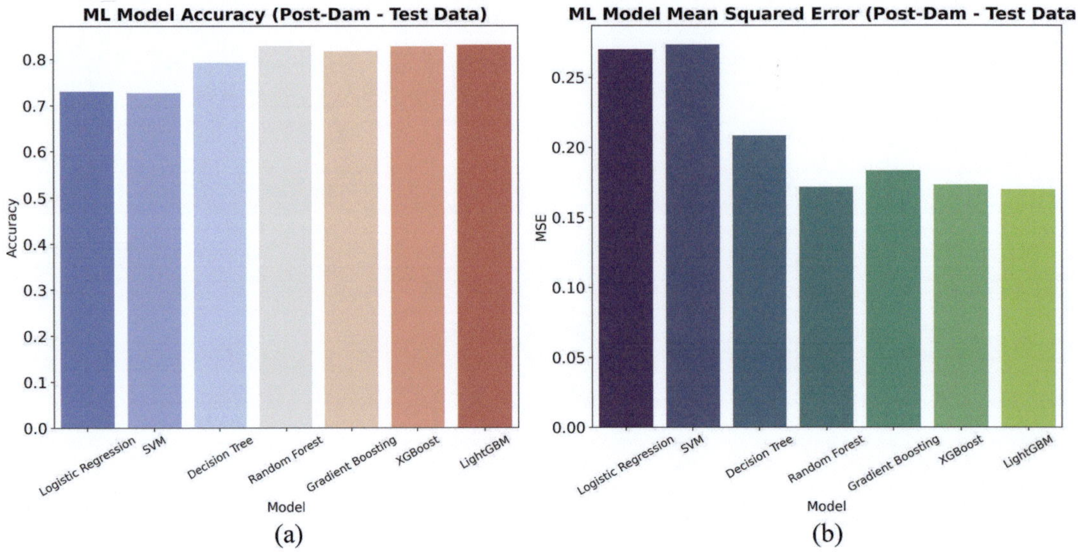

Fig. 8.6 Performance of ML models in assessing wetland health conditions during post-dam conditions with **a** accuracy, **b** MSE values

generalization to the test dataset. The increased error values in these models highlight their limitations in capturing the nonlinear complexities of wetland degradation post-dam construction. Conversely, RF and XGB exhibit significantly lower MSE values, suggesting their superior capability in handling intricate wetland processes under anthropogenic modifications. Compared to the pre-dam condition (Fig. 8.3), post-dam ML model performance shows a slight decline in accuracy across all models, emphasizing the increased complexity of wetland degradation patterns post-dam.

The DL models (Fig. 8.7) exhibit superior performance relative to traditional ML models, with SNN and CNN achieving the highest accuracy (> 81%), indicating their strong aptitude to extract complex spatial and temporal patterns in wetland health deterioration. The MSE analysis further supports this observation, as ANN and WDNN exhibit higher error values, suggesting minor over fitting in the learning process. Notably, the post-dam DL models display reduced accuracy and increased MSE compared to their pre-dam counterparts (Fig. 8.4), indicating the increased uncertainty in wetland response to hydrological alterations. Learning curve assessments (Fig. 8.8) reveal that DL models continue to exhibit fluctuations in training and validation accuracy, with pronounced instability in ANN and WDNN, indicative of potential over fitting. The reduced generalization ability observed in post-dam models further supports the requirement for model tuning and enhanced feature assortment strategies to optimize predictive performance in the altered wetland landscape.

8.4.4 Performance Analysis of the Models

After establishing model optimality, a detailed assessment of model performance is crucial to identify the most suitable approach for wetland health prediction under both pre- and post-dam conditions. Evaluating multiple performance metrics ensures an inclusive understanding of model potency and flaw, facilitating the selection of the best-performing model for long-term wetland monitoring. Figures 8.9 and 8.10 present the relative performance of ML as well as DL models depending on multiple evaluation metrics, including accuracy assessment processes like recall, precision, F1-score, ROC, MCC, Kappa coefficient, balanced accuracy, Jaccard score, and log loss.

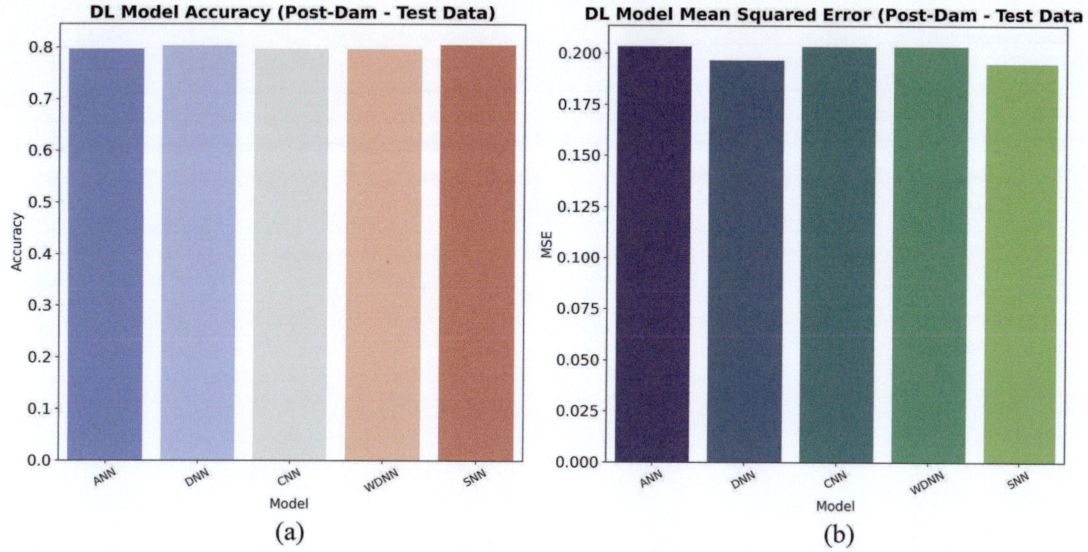

Fig. 8.7 Performance of DL models in assessing wetland health conditions during post-dam conditions with **a** accuracy, **b** MSE values

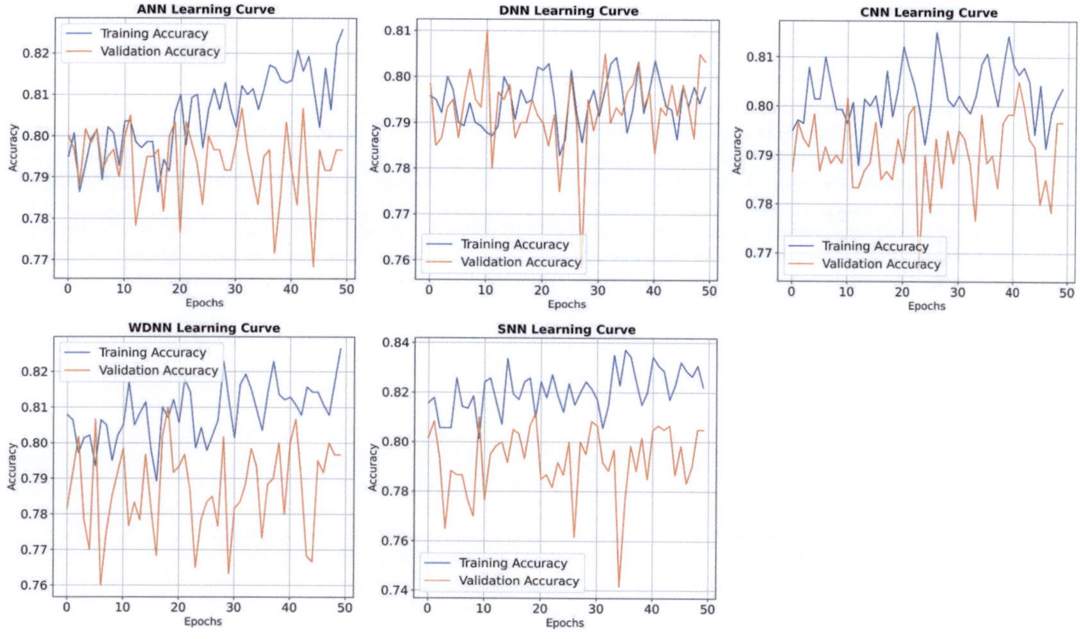

Fig. 8.8 Learning curves of DL models in post-dam conditions

In the pre-dam condition, the best-performing models are XGB, LGBM, and RF, exhibiting correctness values of 0.850, 0.832, and 0.848, in that order (Fig. 8.10). These models also demonstrate high ROC-AUC values above 0.920, indicating strong discriminatory power in classifying wetland health states. Additionally, Cohen's Kappa and MCC values exceed 0.680 for these models, signifying robust agreement between predicted and actual classes. Among DL models, SNN achieves the highest accuracy of 0.827, outperforming ANN and WDNN. The worst-

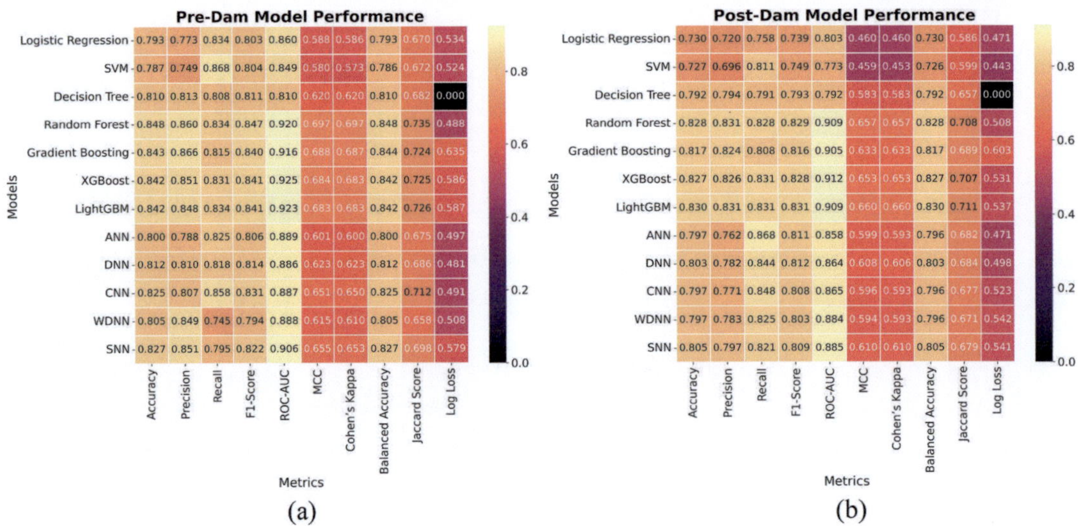

Fig. 8.9 Model performance comparison of wetland health assessment models for **a** pre-dam condition, **b** post-dam condition

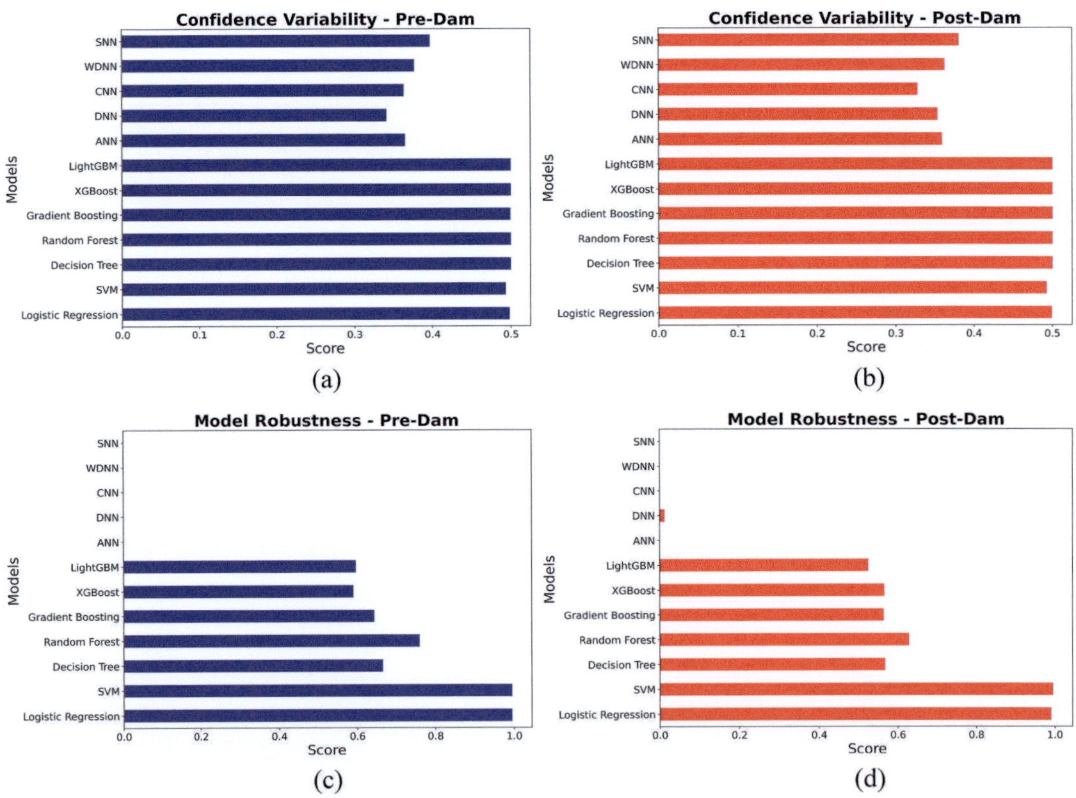

Fig. 8.10 Model confidence variability and robustness in pre- and post-dam conditions. **a** Confidence variability for pre-dam models, **b** confidence variability for post-dam models, **c** model robustness in pre-dam conditions, **d** model robustness in post-dam conditions

performing models include LR and SVM, which register the least accuracy values of 0.793 and 0.787, correspondingly, alongside higher log loss values exceeding 0.524, highlighting their suboptimal performance in wetland health assessment. Under post-dam conditions, model performance exhibits a decline, emphasizing the increased complexity in wetland dynamics due to hydrological modifications (Fig. 8.10). RF, XGB, and LGBM continue to be the top-performing ML models, with accuracy values of 0.828, 0.830, and 0.833, respectively. However, the values are slightly lesser compare to the pre-dam phase, reflecting the added challenges in prediction post-dam construction. The best DL model remains SNN, achieving an accuracy of 0.805, but it also records higher log loss, indicating an increased prediction uncertainty. LR and SVM remain the weakest models, with the least accuracy values of 0.773 and 0.721, along with the poorest ROC-AUC and MCC values, confirming their limited applicability for post-dam wetland health assessment.

Overall, RF, XGB, and LGBM emerge as the most reliable models for both the phases, demonstrating strong predictive ability and stability across multiple performance metrics. DL models, particularly SNN, outperform traditional ML models but exhibit higher log loss, suggesting the need for further fine-tuning. The decline in model performance of post-dam phase signifies the necessity of incorporating advanced feature engineering and hybrid modeling approaches to mitigate prediction uncertainties.

8.4.5 Assessment of Model's Behavior

The assessment of model behavior, including confidence variability, robustness, and inter-model agreement, provides a comprehensive understanding of predictive stability and reliability in wetland health assessment under pre- and post-dam conditions. Evaluating these parameters helps to identify models that maintain predictive consistency while highlighting those that exhibit significant deviations due to data uncertainty, temporal shifts, or algorithmic sensitivity (Figs. 8.10 and 8.11).

Confidence variability, which quantifies the dispersion of model predictions, is a key determinant of prediction stability. In the pre-dam scenario, LR, SVM, RF, GB, and XGB exhibited the highest confidence variability, indicating susceptibility to fluctuations in feature weights. Conversely, DL models such as SNN, WDNN, and CNN showed lower variability, suggesting a more stable representation of wetland health factors (Fig. 8.10). However, post-dam conditions resulted in an increase in confidence variability across all models, particularly for DNN, CNN, and WDNN, emphasizing the influence of altered hydrological and ecological processes. Robustness, defined as a model's resistance to minor perturbations in input data, was highest in LR and DT models in both the phases. RF, GB, and XGB models also demonstrated notable robustness, particularly in the post-dam period, where data complexity and inconsistencies increased. Among DL models, SNN and WDNN had lower robustness, indicating sensitivity to subtle variations in data, particularly under post-dam conditions where wetland fragmentation intensified. The stark contrast between confidence variability and robustness in SVM and RF, which showed high variability but superior robustness, suggests these models adapt to uncertainties while maintaining predictive reliability (Fig. 8.10).

The model agreement matrix provides insights into inter-model dependencies and structural similarities in decision boundaries (Fig. 8.11). In the pre-dam scenario, high agreement (> 0.85) was observed between RF, GB, XGB, and LightGBM, reflecting their shared reliance on hierarchical feature learning. Similarly, SNN and WDNN showed strong agreement (~ 0.92), highlighting their convergent learning pathways in deep architectures. However, ML models such as LR and SVM exhibited lower agreement with tree-based ensembles, indicating fundamental differences in feature selection mechanisms. Post-dam conditions led to an overall reduction in model agreement, particularly among DL models, where the agreement between ANN and CNN dropped from ~ 0.85 to ~ 0.78, suggesting divergence in learned

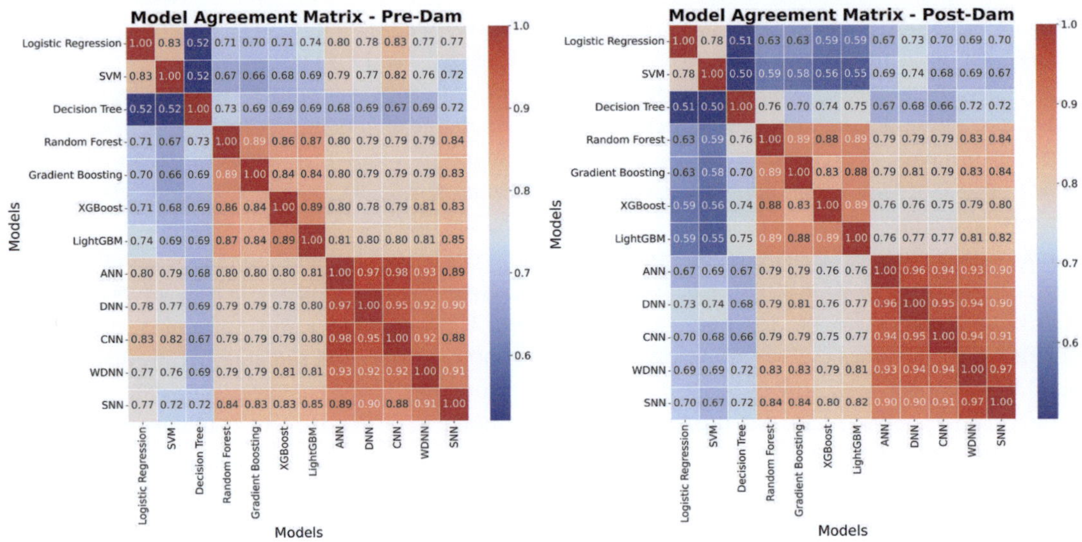

Fig. 8.11 Model agreement matrix in **a** pre-dam and **b** post-dam phases

spatial–temporal patterns under altered wetland conditions. The agreement between RF and XGB remained stable (~ 0.89 pre-dam, ~ 0.88 post-dam), reinforcing the consistency of these tree-based models. Conversely, the decline in agreement for SVM and LR (~ 0.83 pre-dam to ~ 0.78 post-dam) reflects increased sensitivity to hydrological alterations in the post-dam phase.

The findings from confidence variability, robustness, and agreement analyses underscore the influence of computational paradigms on wetland health prediction. RF, GB, and XGB emerged as the most balanced models in reference to confidence, robustness, and agreement, making them ideal for scenario-based wetland health projections. In contrast, DL models, particularly CNN and SNN, exhibited higher adaptability but were more sensitive to post-dam transitions, necessitating further fine-tuning. The trade-off between model stability and sensitivity suggests that ensemble approaches leveraging tree-based models and DL architectures may yield optimal predictive performance for long-term wetland monitoring. These insights highlight the necessity of integrating multiple model architectures to mitigate computational biases while ensuring robustness in wetland degradation assessments. The observed decline in model agreement in post-dam conditions further suggests that feature representations need refinement to capture abrupt shifts.

8.4.6 Spatial Predictions of Wetland Health Conditions

After assessing the optimality and performance of various ML and DL models, the pre-dam raster datasets were integrated into the trained models to predict spatial wetland health conditions. The predicted values ranged from 0 to 1, representing the magnitude of wetland health, where higher values indicated a healthier wetland state, while lower values signified degradation. These continuous predictions were then classified into five distinct wetland health zones—Very good, good, moderate, poor and very poor—following the natural break (Jenks) algorithm to ensure optimal class separation. The results were analyzed spatially to identify directional patterns of wetland health conditions before dam construction (Fig. 8.12).

Spatial Wetland Health Condition Assessment for Pre-dam Conditions

The ML-based wetland health assessment (Fig. 8.13) exhibits substantial variation across

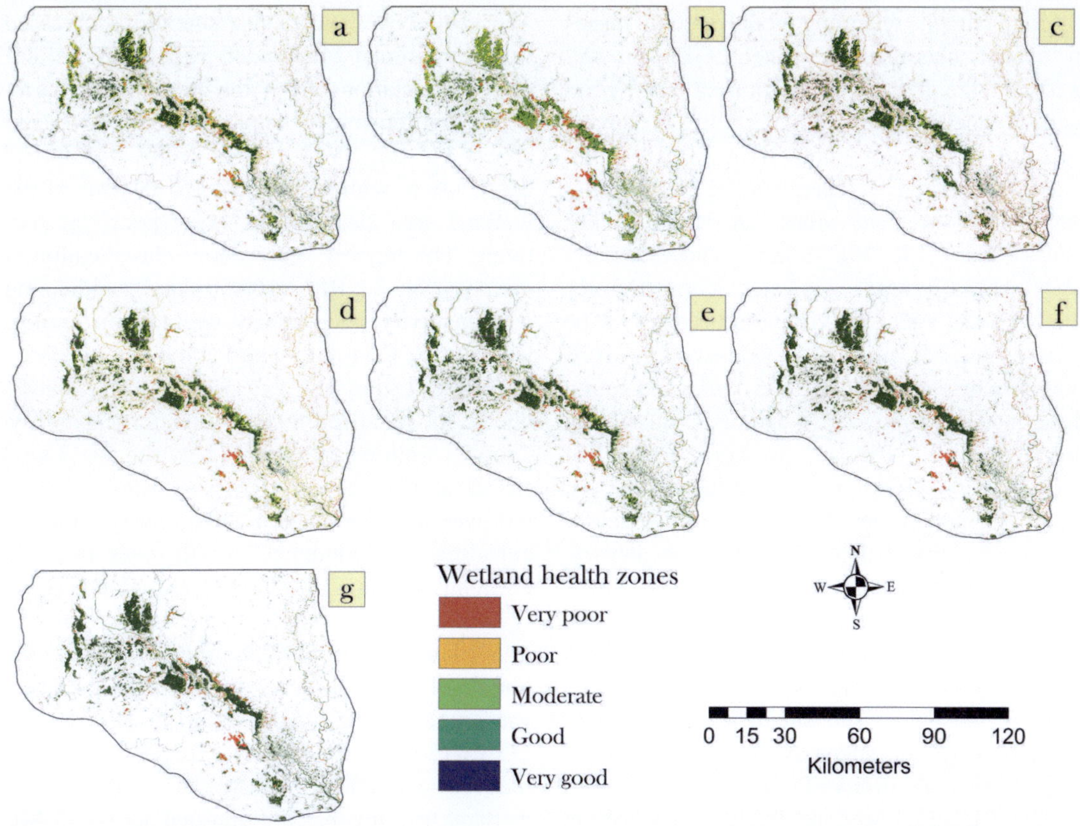

Fig. 8.12 Machine learning-based wetland health classifications before dam construction. **a** LR, **b** SVM, **c** DT, **d** RF, **e** GBM, **f** XGB, **g** LGBM

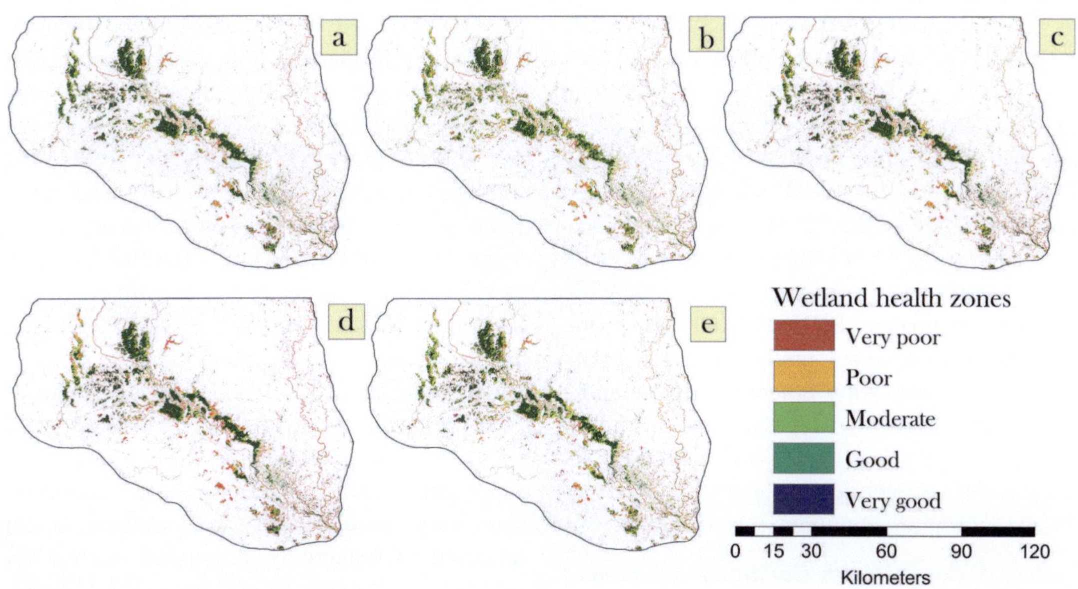

Fig. 8.13 Deep learning-based wetland health classification before dam construction. **a** ANN, **b** DNN, **c** CNN, **d** SNN, **e** WDNN

different models, reflecting model-specific biases in spatial classification. The LGBM model predicted the highest proportion of very good wetland health zones (539.43 km^2), followed by XGB (438.58 km^2) and GBM (392.09 km^2), suggesting effective generalization in detecting high-quality wetland areas. In contrast, DT (297.88 km^2), LR (298.70 km^2), and especially SVM (153.63 km^2) predicted comparatively smaller areas under the very good category. The very poor class was most extensive in XGB (189.46 km^2) and DT (219.06 km^2), indicating these models' conservative stance in classifying degraded zones. RF (118.56 km^2) and GBM (112.40 km^2) reported the smallest very poor areas, suggesting greater precision in distinguishing healthy wetlands. For the poor category, RF showed the highest prediction (242.47 km^2), while GBM (109.12 km^2) and LGBM (74.82 km^2) had the lowest. The moderate class varied notably, with RF again leading (209.15 km^2) and DT presenting the smallest extent (61.02 km^2), underscoring inconsistencies in mid-range classification sensitivity (Table 8.1).

The DL-based wetland health classification (Fig. 8.13) demonstrated distinct spatial variations across architectures, often with a tendency to overestimate degraded conditions. The very poor health category peaked in SNN (320.37 km^2), followed by CNN (251.19 km^2) and WDNN (231.29 km^2), while DNN predicted the least (176.31 km^2). CNN and WDNN showed the highest extent of very good wetland zones (262.97 km^2 and 273.31 km^2, respectively), followed by SNN (242.93 km^2), while DNN again recorded the lowest (226.44 km^2). The good class was best captured by ANN (305.66 km^2), with DNN (261.37 km^2) and CNN (223.50 km^2) showing moderate estimates; SNN (140.29 km^2) and WDNN (132.66 km^2) predicted substantially lower areas, suggesting potential overemphasis on extremes. Moderate wetland zones were best detected by DNN (134.37 km^2) and WDNN (125.38 km^2), reflecting their adaptability to transitional ecological conditions (Table 8.2).

Spatial Wetland Health Condition Assessment for Pre-dam Conditions

The spatial wetland health conditions assessed using traditional ML models (Fig. 8.13) exhibit notable deviations from the pre-dam scenario, with a significant expansion of degraded wetland zones. LR and SVM models depict widespread degradation, with 130.82 and 151.07 km^2 of the wetland area classified as "very poor," respectively. The highest "very poor" classification is observed in LGBM (169.50 km^2), indicating its sensitivity to wetland degradation trends. Conversely, RF (38.81 km^2) exhibits a relatively lower proportion of "very poor" areas, possibly due to its enhanced feature selection capability. "Poor" condition areas range between 64.13 km^2 (XGB) and 129.21 km^2 (RF), suggesting substantial wetland degradation along major riparian corridors. The "moderate" health zones are more concentrated in RF (125.69 km^2) and GBM (77.46 km^2), representing transitional conditions in wetland resilience. The "good" health zone remains limited, with RF predicting the highest value (92.11 km^2), while other models such as LGBM (56.93 km^2) and GBM (69.39 km^2) reflect reduced ecological quality. The "very good" wetland health zones are minimal across all ML models, with the highest being XGB (128.05 km^2), followed by GBM (68.83 km^2) and LGBM (72.10 km^2), reflecting a sharp contrast from the pre-dam landscape (Table 8.3; Fig. 8.14).

The DL-based wetland health classification (Fig. 8.15) further reinforces the degradation trends observed in ML predictions. The "very poor" wetland health category is most prominent in WDNN (135.94 km^2) and SNN (127.94 km^2), highlighting the extensive loss of wetland functionality. CNN, however, shows relatively lower degradation, with 75.35 km^2 classified as "very poor." The "poor" condition zones are highest in CNN (131.78 km^2) and WDNN (107.28 km^2), suggesting these architectures detect extensive degradation in several wetland regions. The "moderate" wetland zones are more balanced, with SNN (90.66 km^2) and DNN (89.89 km^2) capturing transitional health conditions more accurately. The "very good" and "good" wetland health areas have substantially diminished, with ANN predicting the highest "good" zone (113.52 km^2), followed by DNN (103.00 km^2). The "very good"

Table 8.1 Area distribution (km^2) of different wetland health classes predicted using machine learning models

Wetland health zones	Area (Km^2)						
	LR	SVM	DT	RF	GBM	XGB	LGBM
Very poor	174.56	176.82	219.06	118.56	112.40	189.46	112.50
Poor	124.95	122.31	204.61	242.47	109.12	106.50	74.82
Moderate	147.19	139.53	61.02	209.15	137.80	98.68	84.16
Good	224.59	377.72	187.43	165.36	218.59	136.78	159.09
Very good	298.70	153.63	297.88	234.46	392.09	438.58	539.43

Table 8.2 Area distribution (km^2) of different wetland health classes predicted using deep learning models

Wetland health zones	Area (km^2)				
	ANN	DNN	CNN	SNN	WDNN
Very poor	196.80	176.31	251.19	320.37	231.29
Poor	123.07	171.52	129.31	193.45	207.36
Moderate	88.05	134.37	103.04	72.96	125.38
Good	305.66	261.37	223.50	140.29	132.66
Very good	256.41	226.44	262.97	242.93	273.31

class remains marginal across all DL models, peaking in ANN (94.82 km^2), while SNN (43.89 km^2) and WDNN (43.25 km^2) indicate the lowest extent of high-quality wetlands (Table 8.4).

The post-dam wetland assessment reveals an alarming degradation trend compared to pre-dam conditions. The "very poor" category has expanded significantly, although the absolute values remain lower than previously estimated. This expansion is particularly visible in the north western and central wetland zones, where significant hydrological alterations have reduced wetland sustainability. The "poor" category has also increased, reflecting an overall decline in wetland ecosystem functionality. The "moderate" wetland zones appear more fragmented, especially in downstream regions. The most concerning trend is the drastic reduction in the "very good" and "good" wetland health areas. In the pre-dam scenario, RF, GBM, and XGB predicted over 600 km^2 of "very good" wetlands, whereas in the post-dam condition, the highest model estimate is just 128.05 km^2 (XGB). Similarly, DL models such as ANN, DNN, and CNN show significant declines in high-quality wetland areas. This comparative spatial analysis confirms that post-dam conditions have substantially worsened wetland health, with increased fragmentation, hydrological stress, and ecological degradation. North western and central wetland zones have deteriorated most severely, while select south eastern patches still preserve moderate to good ecological conditions. This emphasizes the imperative necessity for targeted protection strategies,

Table 8.3 Area distribution (km^2) of different wetland health classes predicted using machine learning models

Wetland health zones	Area (km^2)						
	LR	SVM	DT	RF	GBM	XGB	LGBM
Very poor	130.82	151.07	130.32	38.81	119.80	121.12	169.50
Poor	89.14	67.33	99.20	129.21	94.53	64.13	73.81
Moderate	80.78	63.57	59.36	125.69	77.46	53.92	57.66
Good	72.85	89.21	78.85	92.11	69.39	62.77	56.93
Very good	56.41	58.82	62.28	44.18	68.83	128.05	72.10

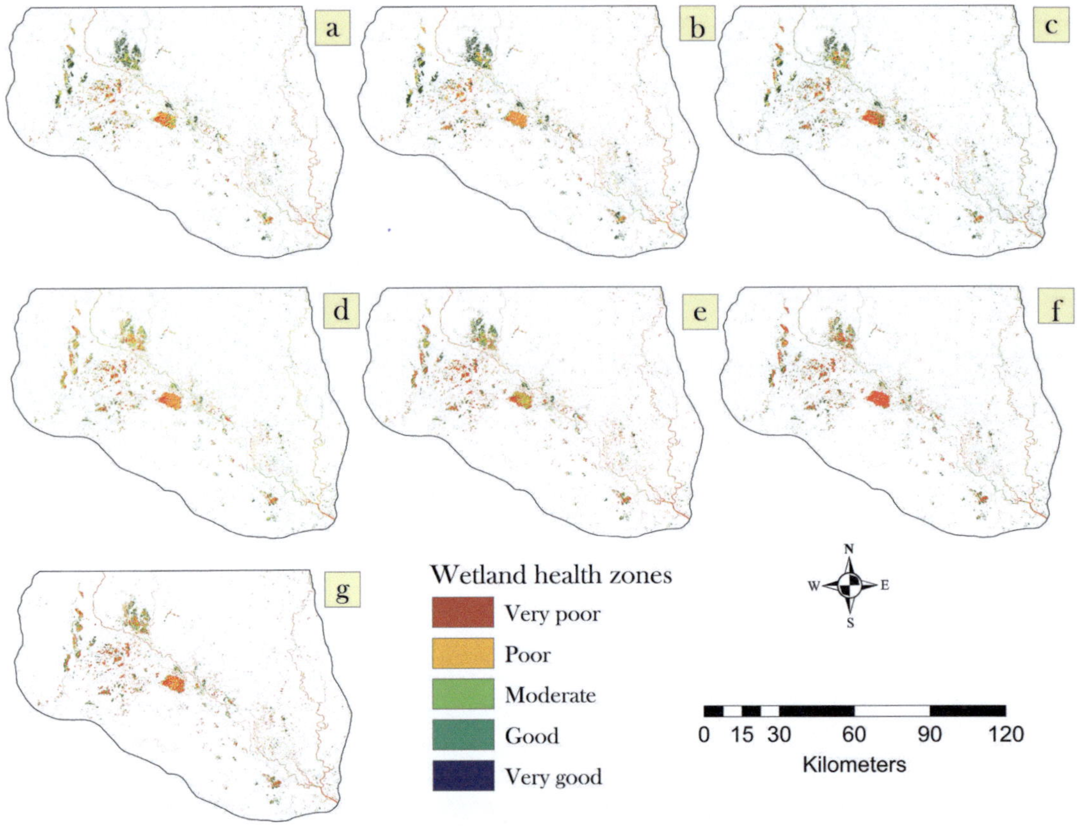

Fig. 8.14 Machine learning-based wetland health classifications after dam construction for **a** LR, **b** SVM, **c** DT, **d** RF, **e** GBM, **f** XGB, **g** LGBM

ecosystem restitution, and hydrological management to alleviate further degradation in post-dam phase.

8.4.7 Utilizing the XAI for Understanding Wetland Health Parameters Behavior

Understanding the primary factors influencing wetland health is critical for formulating conservation and restoration strategies. Figure 8.16 presents the feature importance of different hydrological and ecological indices in determining wetland conditions for both the phases. The variations in feature contributions across ML and DL models provide worthy information of importance into the dynamics of wetland functionality due to dam-induced alterations.

In the pre-dam scenario, the most influential parameters in determining wetland health were HCS, AVDI, and WEI, as evidenced by their high feature importance scores across multiple models. HCS exhibited a strong influence, particularly in RF, XGB, and LGBM, suggesting that well-connected water bodies were crucial for wetland stability before the dam construction. This connectivity facilitated nutrient cycling, sediment transport, and ecosystem resilience, making HCS a vital indicator of conservation priority. Similarly, AVDI had high importance, especially in ANN and XGB models, highlighting the significance of vegetative cover in sustaining wetland stability pre-dam. Vegetation degradation was particularly evident in the northern and central wetland patches, suggesting that hydrological support was essential for preventing vegetation loss and associated biodiversity decline. WEI,

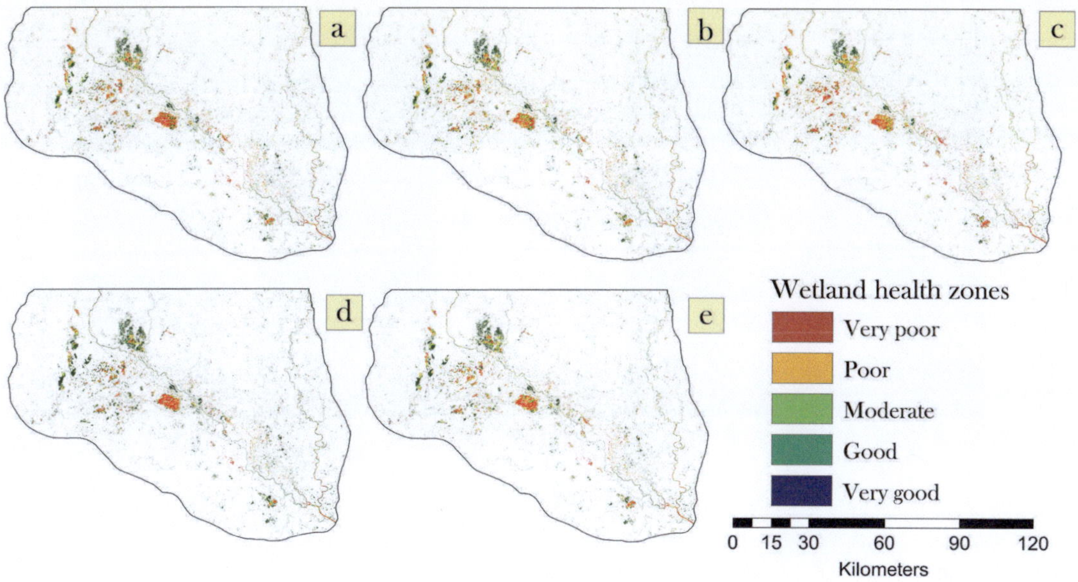

Fig. 8.15 DL-based wetland health classification after dam construction for **a** LR, **b** SVM, **c** DT, **d** RF, **e** GBM, **f** XGB, **g** LGBM

Table 8.4 Area distribution (km^2) of different wetland health classes predicted using machine learning models

Wetland health zones	Area (km^2)				
	ANN	DNN	CNN	SNN	WDNN
Very poor	100.27	87.79	75.35	127.94	135.94
Poor	67.65	87.89	131.78	98.71	107.28
Moderate	53.73	89.89	84.25	90.66	83.35
Good	113.52	103.00	68.36	68.80	60.18
Very good	94.82	61.43	70.26	43.89	43.25

another critical parameter, was found to be highly influential in RF and CNN models, indicating that fluctuations in water availability played a major role in shaping wetland conditions. Regions with high WEI values (~ 3.27, as shown in Fig. 8.1c) were located along the central drainage network and lower reaches, reflecting dynamic hydrological patterns that made these areas particularly sensitive to environmental changes. Other parameters such as Wetland Primary Function Index (WPFI) and Thermal-Water Stress Index (TWSI) exhibited moderate significance, suggesting that terrestrial influences on wetland health were relatively balanced before the dam construction. From a conservation perspective, HCS and WEI are key indicators for maintaining pre-dam hydrological integrity, while AVDI serves as a crucial ecological marker for monitoring vegetation loss. Protecting regions with high hydrological connectivity and stable water presence should be emphasized in conservation planning to warrant long-term wetland ecosystems sustainability.

Post-dam, AESI emerged as the most dominant factor, particularly in ANN and XGB models, indicating that human-induced stressors have become the primary drivers of wetland degradation. The significant increase in AESI values in the central and south western wetland zones suggests that these areas are experiencing severe ecological instability due to hydrological modifications, pollution, and habitat fragmentation. This shift underscores the increased vulnerability of wetlands to anthropogenic alterations, making AESI a critical parameter for wetland restoration strategies. Another significant post-dam indicator is AVDI, which remains

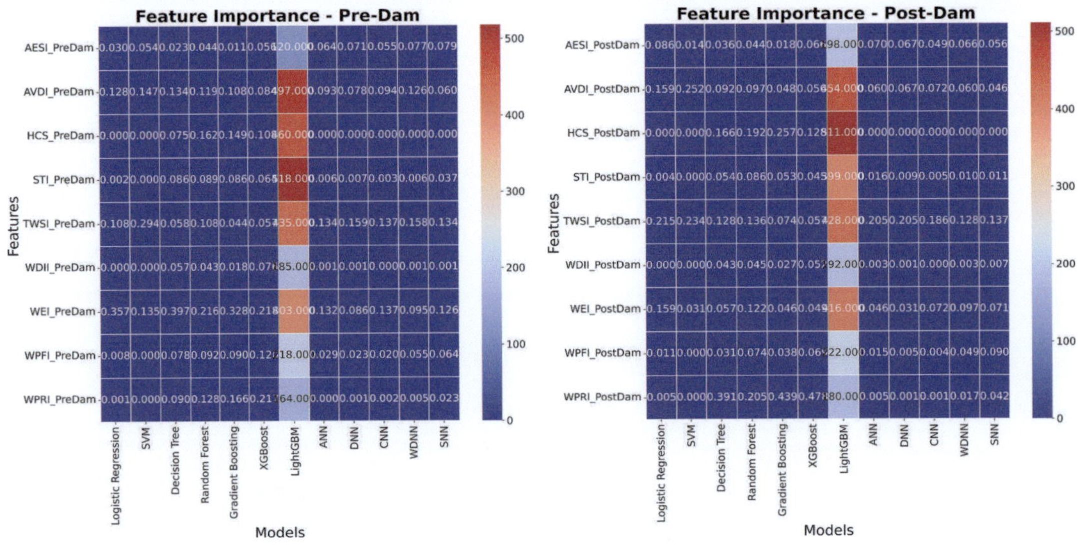

Fig. 8.16 Feature importance analysis for pre-dam and post-dam wetland conditions

highly relevant across multiple models, particularly in LGBM and CNN. However, compared to the pre-dam condition, AVDI values have increased substantially in the northern and central wetland patches, reflecting greater vegetation loss due to hydrological isolation and declining water quality. This pattern highlights the urgent need for vegetation restoration programs, particularly in regions experiencing intensified ecological stress. The importance of WEI has also increased post-dam, indicating that water stability has become more erratic, particularly in areas with fluctuating water availability. This suggests that wetlands are now more susceptible to periodic drying, requiring targeted water management interventions to restore natural hydrological functions. Additionally, TWSI and Water Degradation Intensity Index (WDII) have gained greater importance, suggesting that thermal stress and water degradation have intensified, particularly in the south western and peripheral wetland zones. From a restoration perspective, AESI, AVDI, and WEI must be prioritized in intervention strategies. Addressing anthropogenic stress through pollution control, habitat restoration, and sustainable water management is essential for mitigating further wetland degradation. Furthermore, targeting areas with high AVDI and WEI values for vegetation reforestation and water stability enhancement can improve overall wetland resilience.

A comparison between pre-dam and post-dam feature importance trends reveals a fundamental shift from hydrological stability-driven wetland health to stress-induced degradation mechanisms. While HCS and WEI were the dominant conservation indicators pre-dam, post-dam wetland health is largely governed by AESI, AVDI, and TWSI, reflecting greater human-induced stress and increased hydrological instability. This transition highlights long-term consequences of damming on wetland functionality, highlighting the requirement for integrated conservation and restoration strategies. For wetland conservation, it is essential to preserve hydrological connectivity and maintain stable water regimes to sustain pre-dam ecological functions. This involves restoring natural flow pathways, preventing excessive water diversion, and ensuring adequate sediment transport to maintain habitat integrity. For wetland restoration, priority should be given to regions experiencing severe anthropogenic stress and vegetation loss, particularly areas with high AESI and AVDI values. Restoration efforts should focus on reducing pollution load, enhancing water retention capacity, and implementing adaptive land use management to mitigate further degradation. Additionally, thermal stress reduction strategies, such as increasing vegetation cover and controlling water temperature fluctuations, can significantly improve post-dam wetland resilience. Overall, management strategies must be tailored to both pre-dam conservation needs and post-dam restoration challenges, ensuring a holistic and adaptive approach to wetland sustainability. Figure 8.16 underscores the critical need for long-term monitoring and data-driven decision-making to support evidence-based wetland conservation and restoration efforts.

8.5 Discussion

The assessment of wetland health conditions using ML and DL models provides a comprehensive understanding of the spatial and ecological transformations that have occurred due to dam-induced hydrological modifications. Wetlands, being highly sensitive ecosystems, exhibit significant shifts in their ecological and hydrological characteristics when subjected to external stressors, including alterations in water availability, sedimentation, and vegetation structure. By analyzing pre- and post-dam wetland conditions, this study highlights the extent of wetland degradation through quantitative spatial classification, feature importance analysis, and spatial distribution of wetland health indicators. The integration of traditional ML and advanced DL models has enabled a hierarchical examination of wetland health zones, revealing clear distinctions between stable, transitional, and degraded regions. Pre-dam conditions were characterized by more extensive "very good" and "good" wetland zones, where high hydrological connectivity (HCS) and balanced water availability (WEI) played a crucial role in sustaining wetland functionality. However, post-dam conditions have witnessed a substantial reduction in wetland resilience, with a marked increase in degraded zones, particularly in areas experiencing prolonged water stress, vegetation loss, and pollution accumulation. These observations underscore the critical need for robust computational models to assess wetland health under changing hydrological regimes and provide a basis for targeted conservation and restoration strategies.

The evaluation of ML and DL models for wetland health assessment revealed substantial differences in predictive accuracy, robustness, and generalization across pre- and post-dam conditions. Traditional ML models, including LR, SVM, DT, RF, GB, XGB, and LGBM, exhibited distinct performance patterns, with tree-based ensemble methods such as RF, GB, and XGB achieving superior accuracy (> 0.82) and lower MSE in pre-dam conditions. The efficacy of tree-based models aligns with Talebi et al. (2022) and Bharti et al. (2021), who highlighted RF and XGB as robust classifiers for spatial prediction owing to their capability to manage high-dimensional datasets. Conversely, LR and SVM, despite their interpretability, exhibited suboptimal performance (accuracy < 0.79), corroborating findings from Niaz et al. (2021), who demonstrated the limitations of logistic regression in capturing complex spatial patterns. Post-dam, ML models experienced a decline in predictive performance, with accuracy reductions of ~ 2–4%, indicating increased environmental complexity. XGB and LGBM retained the highest predictive power, reflecting their adaptability to dynamic wetland conditions, as demonstrated in Kavzoglu and Teke (2022). SVM, in contrast, exhibited increased MSE, reinforcing the findings of Wang (2022) and Sahu and Pandey (2023), who noted the sensitivity of SVM to nonlinear environmental variations. Deep learning models, particularly CNN and SNN, demonstrated superior accuracy (> 0.84) compared to ANN and WDNN, unswerving with findings from Guha et al. (2022) and Yao et al. (2022), where CNN outperformed conventional ML in spatial susceptibility mapping. However, the post-dam scenario introduced challenges, with validation accuracy fluctuations indicating potential overfitting in WDNN and ANN, similar to findings by Liu et al. (2022). Learning curves highlighted SNN's superior convergence, reinforcing its robustness for complex environmental prediction, as supported by Zhao et al. (2025). Despite higher computational demands, DL models demonstrated greater adaptability in capturing wetland degradation patterns, underscoring their potential for long-term environmental monitoring (Alatarany et al. 2023). The observed decline in post-dam accuracy (~ 2–3%) suggests the necessity of hybrid model optimization, integrating ML's interpretability with DL's feature extraction, as proposed by Wei et al. (2021). These findings advocate for an ensemble-based wetland monitoring framework, leveraging the strengths of both ML and DL to enhance predictive resilience in anthropogenically altered landscapes.

The spatial assessment of wetland health conditions using ML and DL models reveals significant shifts in wetland degradation patterns, highlighting the profound impact of dam-induced hydrological modifications. The pre-dam condition exhibits a more balanced distribution of wetland health zones, with 865.57 km^2 of "very good" areas predicted by LGBM, while RF and XGB classified over 600 km^2 in this category (Table 8.1). This suggests a relatively stable wetland ecosystem where hydrological connectivity and vegetation cover were sustained. However, post-dam, the extent of "very good" zones has drastically declined, with LGBM predicting only 261.03 km^2, and RF recording an even lower 160.45 km^2 (Table 8.3). This steep reduction indicates a collapse in wetland resilience due to disrupted hydrodynamics, fragmentation, and altered sediment deposition patterns. The "very poor" zones, which occupied only 347.48 km^2 in pre-dam conditions (DT model), expanded to 613.65 km^2 post-dam (LGBM), reflecting intensified habitat degradation, increased drying trends, and loss of vegetation support systems. This transformation aligns with feature importance analysis, where HCS was a dominant factor pre-dam, reinforcing the role of uninterrupted water exchange in wetland stability. However, post-dam, AESI and AVDI gained prominence, indicating that the wetland ecosystem is now primarily influenced by stress-induced fragmentation and vegetative loss rather than natural hydrological factors. The drastic increase in AVDI values (~ 149.9 post-dam vs. ~ 87.25 pre-dam) suggests that sustained hydrological deficits have triggered a cascading effect, leading to vegetation stress, reduced soil moisture, and declining habitat conditions. This aligns with the spatial distribution patterns, where central and northern wetland patches, previously categorized under "good" and "very good" zones, have transitioned into "poor" and "very poor" categories post-dam, indicating widespread degradation of key wetland areas. The feature importance analysis further elucidates the cause-effect relation between wetland degradation and dominant ecological stressors. Pre-dam, HCS, AVDI, and WEI collectively governed wetland stability, suggesting that a well-connected hydrological system supported ecosystem resilience by sustaining vegetation and balancing water availability. However, post-dam, AESI emerges as the dominant factor, highlighting the role of anthropogenic stressors in accelerating wetland degradation. The shift in feature dominance suggests a fundamental transition from hydrologically regulated stability to ecologically driven instability, where wetlands are increasingly governed by human-induced stress factors such as altered water retention, sedimentation changes, and pollution. The worsening of TWSI (~ 21.82 pre-dam to ~ 16.2 post-dam) in downstream zones indicates that thermal stress has intensified due to reduced buffering capacity, leading to amplified evapotranspiration rates. Similarly, WPFI, which exhibited higher values pre-dam (~ 0.187), has become more spatially fragmented post-dam (~ 0.197), reinforcing the idea that hydrological inconsistency has increased, making wetlands more susceptible to seasonal fluctuations. The north western and south western regions, previously exhibiting stable wetland conditions, now show elevated WDII (~ 0.015 post-dam vs. ~ 0.09 pre-dam), suggesting that water degradation has intensified due to prolonged retention of pollutants and reduced natural flushing mechanisms. The strong correlation between AESI, AVDI, and WPRI in post-dam conditions indicates that wetland degradation is now primarily driven by hydrological fragmentation, pollution accumulation, and thermal stress, making restoration efforts increasingly challenging. These findings bring to light the imperative need for dedicated conservation strategies, prioritizing restoration of hydrological connectivity, vegetation reinforcement, and pollutant regulation to mitigate further degradation. Adaptive water management interventions are critical to reversing the trend of wetland fragmentation and restoring ecosystem equilibrium.

The findings from the wetland health assessment highlight a significant deterioration in wetland stability and ecosystem integrity post-dam, with increased vulnerability to hydrological and anthropogenic stressors. The stark decline in "very good" wetland zones and the expansion of "very poor" and "poor" categories indicate that hydrological fragmentation, prolonged retention of pollutants, and intensified vegetation degradation have collectively contributed to the decline in wetland sustainability. The shift in feature dominance from hydrologically driven stability indicators (HCS, WEI, WPFI) pre-dam to stress-induced degradation markers (AESI, AVDI, TWSI, WPRI) post-dam suggests that wetland resilience has been significantly compromised due to external modifications. This transformation necessitates a strategic approach to conservation, emphasizing the restoration of hydrological connectivity, mitigation of thermal stress, and control of pollutant accumulation to prevent further ecosystem degradation. While ML and DL models effectively capture wetland degradation patterns, their performance is influenced by model selection, data complexity, and feature interactions, requiring further advancements in XAI for improving prediction reliability. The study highlights the urgency of integrating long-term monitoring frameworks with advanced computational modeling to track wetland health trends and inform adaptive management interventions. Moving forward, targeted conservation policies, eco-hydrological restoration initiatives, and sustainable use of land are essential to reversing the negative impacts of dam-induced modifications and certifying the long-term sustainability of wetland ecosystems.

This study presents a comprehensive assessment of wetland health dynamics, integrating multi-sensor remote sensing datasets, machine learning (ML), deep learning (DL), and Explainable AI (XAI) to analyze pre- and post-dam wetland conditions. By developing novel wetland health parameters, including hydrological, ecological, and thermal indices, we quantified spatio-temporal variations in wetland stability and degradation. The findings reveal that pre-dam wetland health was largely governed by hydrological connectivity (HCS: 0.6) and water persistence (WPFI: 0.187), whereas post-dam conditions were dominated by anthropogenic stress (AESI: -11.47) and vegetation degradation (AVDI: 149.9), indicating a transition from hydrological resilience to stress-driven ecosystem deterioration. Model evaluation demonstrated higher accuracy in tree-based ML models (RF: 0.848, XGB: 0.850) and CNN/SNN DL models (> 0.82) in pre-dam conditions, but performance declined post-dam, with increased uncertainty and degradation detection errors. The "very poor" wetland class expanded significantly, reaching 613.65 km^2 (LGBM, post-dam) compared to 347.48 km^2 (DT, pre-dam), underscoring severe ecological stress. The originality of this work lies in integrating traditional and advanced models with XAI to interpret key environmental drivers, facilitating a more explainable and data-driven method for wetland monitoring. However, limitations include potential errors in multi-source data fusion, model sensitivity to extreme events, and the need for higher temporal resolution datasets. Future research should focus on hybrid ensemble modeling, real-time wetland monitoring using UAVs, and predictive simulations of climate change impacts on wetland functionality. The outcomes present a scientific foundation for wetland conservation strategies, advocating for hydrological restoration, pollution control, and adaptive land use planning to mitigate human-induced wetland degradation and ensure long-term ecological sustainability.

References

Abdelmajeed AYA, Albert-Saiz M, Rastogi A, Juszczak R (2023) Cloud-based remote sensing for wetland monitoring—a review. Remote Sens 15(6):1660

Ahmed KR, Akter S, Marandi A, Schüth C (2021) A simple and robust wetland classification approach by using optical indices, unsupervised and supervised machine learning algorithms. Remote Sens Appl Soc Environ 23:100569

Alatrany AS, Khan W, Hussain A, Al-Jumeily D, Alzheimer's Disease Neuroimaging Initiative (2023) Wide and deep learning based approaches for classification of Alzheimer's disease using genome-wide association studies. Plos One 18(5):e0283712

Aslam RW, Shu H, Naz I, Quddoos A, Yaseen A, Gulshad K, Alarifi SS (2024) Machine learning-based wetland vulnerability assessment in the sindh province ramsar site using remote sensing data. Remote Sens 16(5):928

Bharti JP, Mishra P, Moorthy U, Sathishkumar VE, Cho Y, Samui P (2021) Slope stability analysis using Rf, gbm, cart, bt and xgboost. Geotech Geol Eng 39:3741–3752

Chen L, Jin Z, Michishita R, Cai J, Yue T, Chen B, Xu B (2014) Dynamic monitoring of wetland cover changes using time-series remote sensing imagery. Eco Inform 24:17–26

Darabi H, Haghighi AT, Rahmati O, Shahrood AJ, Rouzbeh S, Pradhan B, Bui DT (2021) A hybridized model based on neural network and swarm intelligence-grey wolf algorithm for spatial prediction of urban flood-inundation. J Hydrol 603:126854

Das S, Adhikary PP, Shit PK, Bera B (2022) Urban wetland fragmentation and ecosystem service assessment using integrated machine learning algorithm and spatial landscape analysis. Geocarto Int 37(25):7800–7818

Demarquet Q, Rapinel S, Dufour S, Hubert-Moy L (2023) Long-term wetland monitoring using the landsat archive: a review. Remote Sensing 15(3):820

Ekmekcioğlu Ö, Koc K (2022) Explainable step-wise binary classification for the susceptibility assessment of geo-hydrological hazards. CATENA 216:106379

Ghobadi Y, Pradhan B, Kabiri K, Pirasteh S, Shafri HZM, Sayyad GA (2012) Use of multi-temporal remote sensing data and GIS for wetland change monitoring and degradation. In: 2012 IEEE Colloquium on humanities, science and engineering (CHUSER). IEEE, pp 103–108

Gonzalez-Perez A, Abd-Elrahman A, Wilkinson B, Johnson DJ, Carthy RR (2022) Deep and machine learning image classification of coastal wetlands using unpiloted aircraft system multispectral images and lidar datasets. Remote Sens 14(16):3937

Gu T, Li J, Wang M, Duan P (2022) Landslide susceptibility assessment in Zhenxiong County of China based on geographically weighted logistic regression model. Geocarto Int 37(17):4952–4973

Guha S, Jana RK, Sanyal MK (2022) Artificial neural network approaches for disaster management: a literature review. Int J Disaster Risk Reduct 81:103276

Günen MA (2022) Performance comparison of deep learning and machine learning methods in determining wetland water areas using EuroSAT dataset. Environ Sci Pollut Res 29(14):21092–21106

Hasani SS, Mojtahedi A, Reshadi MAM (2021) Development of a novel fuzzy logic-based wetland health assessment approach for the management of freshwater wetland ecosystems. Wetlands 41:1–25

Hosseiny B, Mahdianpari M, Brisco B, Mohammadimanesh F, Salehi B (2021) WetNet: a spatial–temporal ensemble deep learning model for wetland classification using Sentinel-1 and Sentinel-2. IEEE Trans Geosci Remote Sens 60:1–14

Jafarzadeh H, Mahdianpari M, Gill EW, Brisco B, Mohammadimanesh F (2022) Remote sensing and machine learning tools to support wetland monitoring: a meta-analysis of three decades of research. Remote Sens 14(23):6104

Jamali A, Mahdianpari M, Brisco B, Granger J, Mohammadimanesh F, Salehi B (2021) Wetland mapping using multi-spectral satellite imagery and deep convolutional neural networks: a case study in Newfoundland and Labrador, Canada. Can J Remote Sens 47(2):243–260

Jianping LI, Bai ZHANG, Ling Z, Zongming WANG, Kaishan SONG (2007) Current status and prospect of researches on wetland monitoring based on remote sensing. Prog Geogr 26(1):33–43

Kaplan G, Avdan U (2018) Monthly analysis of wetlands dynamics using remote sensing data. ISPRS Int J Geo Inf 7(10):411

Kavzoglu T, Teke A (2022) Advanced hyperparameter optimization for improved spatial prediction of shallow landslides using extreme gradient boosting (XGBoost). Bull Eng Geol Env 81(5):201

Lee TM, Yeh HC (2009) Applying remote sensing techniques to monitor shifting wetland vegetation: a case study of Danshui River estuary mangrove communities, Taiwan. Ecol Eng 35(4):487–496

Li Q, Wang D, Liu F, Yu J, Jia Z (2024) LightGBM hybrid model based DEM correction for forested areas. PLoS ONE 19(10):e0309025

Liu R, Yang X, Xu C, Wei L, Zeng X (2022) Comparative study of convolutional neural network and conventional machine learning methods for landslide susceptibility mapping. Remote Sens 14(2):321

Madasu A, Rao VA (2019) Effectiveness of self normalizing neural networks for text classification. In: International conference on computational linguistics and intelligent text processing. Springer, Cham, pp 412–423

Mahdavi S, Salehi B, Granger J, Amani M, Brisco B, Huang W (2018) Remote sensing for wetland classification: a comprehensive review. Gisci Remote Sens 55(5):623–658

Martínez Prentice R, VillosladaPeciña M, Ward RD, Bergamo TF, Joyce CB, Sepp K (2021) Machine learning classification and accuracy assessment from high-resolution images of coastal wetlands. Remote Sens 13(18):3669

Mondal I, Naskar PK, Alsulamy S, Jose F, Hossain SA, Mohammad L et al (2024) Habitat quality and degradation change analysis for the Sundarbans mangrove forest using invest habitat quality model and machine learning. Environ Develop Sustain 1–26

Nafouanti MB, Li J, Mustapha NA, Uwamungu P, Al-Alimi D (2021) Prediction on the fluoride contamination in groundwater at the Datong Basin, Northern China: comparison of random forest, logistic regression and artificial neural network. Appl Geochem 132:105054

Niaz R, Zhang X, Iqbal N, Almazah MM, Hussain T, Hussain I (2021) Logistic regression analysis for spatial patterns of drought persistence. Complexity 2021(1):3724919

Orimoloye IR, Mazinyo SP, Kalumba AM, Nel W, Adigun AI, Ololade OO (2019) Wetland shift monitoring using remote sensing and GIS techniques: landscape dynamics and its implications on Isimangaliso Wetland Park, South Africa. Earth Sci Inf 12:553–563

Ozesmi SL, Bauer ME (2002) Satellite remote sensing of wetlands. Wetlands Ecol Manage 10:381–402

Pham QB, Chandra Pal S, Chakrabortty R, Saha A, Janizadeh S, Ahmadi K et al (2022). Predicting landslide susceptibility based on decision tree machine learning models under climate and land use changes. Geocarto Int 37(25):7881–7907

Rawat M, Pandey A, Gupta PK, Yadav B, Patel JG (2025) A novel framework for wetland health assessment using hydro-ecological indicators and landscape metrics. Model Earth Syst Environ 11(3):167

Saber M, Boulmaiz T, Guermoui M, Abdrabo KI, Kantoush SA, Sumi T et al (2022) Examining LightGBM and CatBoost models for wadi flash flood susceptibility prediction. Geocarto Int 37(25):7462–7487

Sahani N, Ghosh T (2021) GIS-based spatial prediction of recreational trail susceptibility in protected area of Sikkim Himalaya using logistic regression, decision tree and random forest model. Eco Inform 64:101352

Sahin EK (2022) Comparative analysis of gradient boosting algorithms for landslide susceptibility mapping. Geocarto Int 37(9):2441–2465

Sahu SK, Pandey M (2023) An optimal hybrid multiclass SVM for plant leaf disease detection using spatial Fuzzy C-Means model. Expert Syst Appl 214:118989

Shafi A, Chen S, Waleed M, Sajjad M (2023) Leveraging machine learning and remote sensing to monitor long-term spatial-temporal wetland changes: towards a national RAMSAR inventory in Pakistan. Appl Geogr 151:102868

Shuman CS, Ambrose RF (2003) A comparison of remote sensing and ground-based methods for monitoring wetland restoration success. Restor Ecol 11(3):325–333

Singh S, Goyal MK, Saikumar E (2024) Assessing climate vulnerability of Ramsar wetlands through CMIP6 projections. Water Resour Manage 38(4):1381–1395

Steinbach S, Hentschel E, Hentze K, Rienow A, Umulisa V, Zwart SJ, Nelson A (2023) Automatization and evaluation of a remote sensing-based indicator for wetland health assessment in East Africa on national and local scales. Eco Inform 75:102032

Talebi H, Peeters LJ, Otto A, Tolosana-Delgado R (2022) A truly spatial random forests algorithm for geoscience data analysis and modelling. Math Geosci 54(1):1–22

Van Le H, Hoang DA, Tran CT, Nguyen PQ, Tran VHT, Hoang ND et al (2021) A new approach of deep neural computing for spatial prediction of wildfire danger at tropical climate areas. Ecol Inform 63:101300

Wang Q (2022) Support vector machine algorithm in machine learning. In: 2022 IEEE International conference on artificial intelligence and computer applications (ICAICA). IEEE, pp 750–756

Wei X, Zhang L, Luo J, Liu D (2021) A hybrid framework integrating physical model and convolutional neural network for regional landslide susceptibility mapping. Nat Hazards 109:471–497

Yadav PK, Jha P, Joy MS, Bansal T (2024) Ecosystem health assessment of East Kolkata Wetlands, India: implications for environmental sustainability. J Environ Manage 366:121809

Yang W, Fu B, Li S, Lao Z, Deng T, He W et al (2023) Monitoring multi-water quality of internationally important karst wetland through deep learning, multi-sensor and multi-platform remote sensing images: a case study of Guilin, China. Ecol Indicators 154:110755

Yao J, Qin S, Qiao S, Liu X, Zhang L, Chen J (2022) Application of a two-step sampling strategy based on deep neural network for landslide susceptibility mapping. Bull Eng Geol Env 81(4):148

Yousefzadeh M, Hosseini SA, Farnaghi M (2021) Spatiotemporally explicit earthquake prediction using deep neural network. Soil Dyn Earthq Eng 144:106663

Zhao Y, Yang L, Pan H, Li Y, Shao Y, Li J, Xie X (2025) Spatio-temporal prediction of groundwater vulnerability based on CNN-LSTM model with self-attention mechanism: a case study in Hetao Plain, northern China. J Environ Sci 153:128–142

Zheng JY, Hao YY, Wang YC, Zhou SQ, Wu WB, Yuan Q et al (2022) Coastal wetland vegetation classification using pixel-based, object-based and deep learning methods based on RGB-UAV. Land 11(11):2039

9 Modeling Restoration of Wetland

Abstract

This chapter involved showing some ways forward for wetland restoration and protection considering their ecological and societal importance. In first part, this chapter tried to model restoration sites majorly thrusting on specific conditions on which the existing wetlands are surviving. Supervised classification approaches using maximum likelihood classifier (MLC) and support vector machine (SVM) were adopted for restorable site identification. Both the models showed restorable sites are located within the major wetland tract of the river basin. Most of the cases the restorable sites are very contiguous to the existing wetland. The used models exhibited that about 41–122 km^2 area could be restored. Moreover to do this, the study recommended releasing ecological flow from the dam for regular inundation and revival of tie channels and floodplain fluvial passages. Since, damming was identified as a one of the major causes for flow reduction in river and wetland loss, so ecological maintenance of flow downstream is very crucial.

Keywords

Wetland loss · Wetland heath deterioration · Restoration · Modeling restoration sites and recommendations

9.1 Concept and Utility of Restoration

Wetland restoration is an approach of re-establishment (rebuilding a former wetland), redevelopment, rehabilitation (repairing functions), of degraded or lost wetland's physical, chemical, or biological functions in its natural conditions (US EPA 2007). This process includes restoring hydrological conditions, restoring soil, reintroducing wildlife, replanting native vegetation, and managing stresses like water overuse or pollution (Zedler 2000). Hydrological restoration involves restoring natural water flow removing longitudinal barriers likes dams, barrages, dykes, etc. and lateral barriers like embankments, rejuvenation, and re-connection of tie channels (Meng et al. 2020). Vegetation restoration involves implanting native wetland plants and removing invasive exotic species. Soil restoration involves de-silting wetland bed where necessary, create ambience for re-establishing soil profile, improvement of soil drainage, etc. Faunal restoration involves reintroduction of native wildlife species better suit to wetland environment. Wetland restoration mechanism could offer significant benefits, including enhancement of biodiversity, improvement of water quality, flood and storm control, wild life support, increase of environment purification, deceleration of soil loss, enhancement of water security, increase of climate resilience, socioeco-

S. Pal et al., *Damming and Wetland Eco-hydrology*, GIScience and Geo-environmental Modelling,
https://doi.org/10.1007/978-3-032-11729-8_9

nomic benefits (Mitsch et al. 2015) extending wide range of ecosystem services, etc.

Over the past few decades, global efforts to restore wetlands have gained momentum due to the alarming decline in the ecosystem services which these habitats used to provide. It is estimated that between 64 and 71% of the world's wetlands have been lost since the beginning of the twentieth century, primarily due to conversion wetland for agriculture and development purposes. This degradation has significantly impacted the provisioning, regulating, and supporting services offered by wetland ecosystems (Ausseil et al. 2008; Tomscha et al. 2021). Recognizing the urgency of reversing this trend, the United Nations declared 2021–2030 as the Decade on Ecosystem Restoration in March 2019 (Rodrigo 2021). In this context, wetland restoration should be recognized as a high-priority action within environmental management and policy frameworks.

Climate change and human interventions issues are very crucially withstanding against wetland restoration endeavor. Human activities such as agricultural expansion, efforts to enhance water security by modifying rivers and wetlands, industrial development, and various forms of pollution are among the primary drivers of wetland degradation in floodplain areas. These anthropogenic pressures, as noted by Sievers et al. (2017), pose significant challenges to the successful restoration and long-term sustainability of wetland ecosystems. People are not ready to forego the already reclaimed lands. No acceptable measures were taken to restrict pollutants to the river and wetland. Gradual accretion and wetland isolation strongly detect the rapid spacing change in wetland quality in the agricultural dominated floodplain area (Karim et al. 2015). Amidst these adverse ecological ambiences, it is really challenging to adopt and implement restoration measures. Besides all, absorbing all the stresses, some wetlands are still surviving. These would be inspiration toward the efforts of wetland restoration. Scientific investigation of physical and socio-ecological ambience of the exiting wetlands and searching the same environment in other areas could be effective step for identifying the appropriate sites for wetland restoration. Therefore, based on this thinking, in present study, the endeavor was made to model and identify the restoration sites. Since the earlier chapters clearly noted the hydrological modifications in the primary feeding river as one of the major drivers of wetland loss and degradation, so restoration of river flow would be the best step for restoration efforts. But considering prevailing socio-ecological condition, it may not be possible to restore channel flow entirely. However, release of ecological flow through river with bilateral dialogue and identification of potential wetland restoration sites could make a sustainable wetland condition, and hence, the work was undertaken. Spieles (2022) clearly reported that the benefits of constructed wetlands and restored wetland citing different examples across the world. For instance, literatures showed 77–99% retention of BOD, COD, TSS, TP, TN; enhancement of significant amount of carbon sequestration, supporting > 600 aquatic species, supporting regional water and nutrient cycling, etc. In case of restored natural wetland, 36% greater provisioning services were retrieved than degraded wetland, improved water and nutrient cycling, ecosystem connectivity; 29–99% retention of BOD, COD, TSS, TP, TN; supported thousands of plant species, etc. (Mitsch et al. 1998; Dai et al. 2022). So, these success evidences are the inspiration for adopting restoration strategies.

9.2 Materials

To identify potential sites suitable for wetland restoration, six key spatial parameters were considered: (1) proximity of wetland to agricultural land, (2) drainage density, (3) slope, (4) elevation, (5) proximity of wetland to built-up areas, and (6) surface curvature. These factors collectively influence the ecological and hydrological feasibility of restoration interventions. SRTM DEM data were taken for preparation of first four parameters. Proximity to agriculture land and built area were developed from LANDSAT images of post-monsoon seasons, 2024. Training data were created from existing wetland map of post-monsoon, 2024 and field investigation. Total 124 training sites were taken, and out of these, 45 were collected from field.

9.3 Methods

9.3.1 Modeling Wetland Restoration Sites

9.3.1.1 Justification for Selection and Preparation of Data Layers

In floodplain region, elevation is one of the major determinants of water stagnation. Lowly elevated, very mild slope with poor drainage area is potential for water stagnation. Concave part of the floodplain is more susceptible to water stagnation (Peng et al. 2024). Considering these elevation, slope and curvature parameters were taken into account. These were prepared from DEM data. Drainage density particularly higher order drainage density in lower basin often creates ambient conditions for high flood and greater water availability in river proximate floodplain area (Odoh and Nwokeabia 2024). For drainage density development, first drainage boundary and drainage map was developed from DEM data in Arc GIS and Strahler's stream order was executed from the same platform. Only fifth to seventh (highest) order streams were considered and drainage density map was prepared accordingly in the same environment.

As agricultural encroachment is recognized as a significant contributor to wetland degradation in floodplain regions (Johnston and McIntyre 2019), this parameter was incorporated into the present analysis to assess its impact on restoration suitability. Built area development reclaiming wetland is another common incident all across the world (Mao et al. 2018); hence, this factor was also considered here. First maximum likelihood classifier-based land use classification from Landsat image of post-monsoon, 2024 was done. Agriculture and built area components were taken separately for proximity analysis. In both the cases, it was assumed that the area away from agriculture and built area would be potential for wetland restoration.

Figure 9.1 depicts the spatial data layers considered for modeling wetland restoration. Here, it is to be mentioned that discharge induce flood extent could be taken since dam-induced

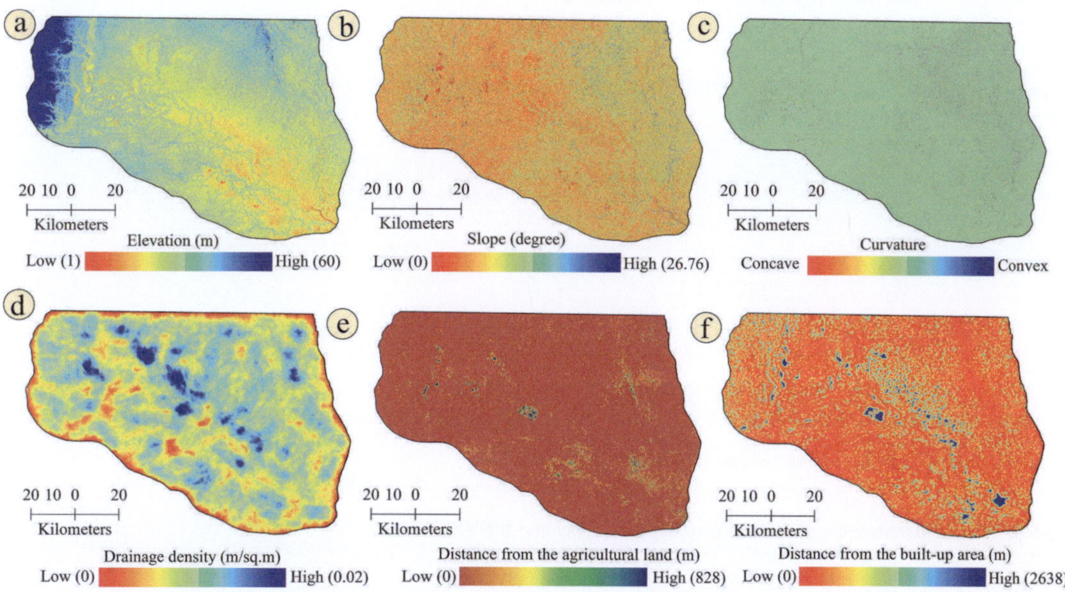

Fig. 9.1 Wetland restoration conditioning parameters taken for the study **a** elevation, **b** slope, **c** curvature, **d** drainage density, **e** distance from the agriculture land, **f** distance from the built area

flood reduction is a crucial factor behind wetland loss (Karim et al. 2020). However, inadequate data for recent time restricted to include this parameter.

9.3.1.2 Methods of Modeling

Total 124 numbers of training sites from existing wetland and 45 numbers from field sites were taken from training for restorable sites identification. It was assumed that the ambience like existing wetland would be potential since the existing wetlands are surviving themselves absorbing all the stresses posed from nature and human activities. Here, stresses were given on the recently lost wetlands for fining out restorable site identification. Therefore, historical wetland maps were also corroborated.

Supervised classification approaches using maximum likelihood classifier (MLC) and support vector machine (SVM) were adopted for restorable site identification.

Since, no restoration mechanism was adopted hence, no success was evaluated anywhere within the study area so far. So, it is difficult to validate the models. But it would be more acceptable if it would be done. Even among the models produced, assessment of outperformance was also required. This is the limitation of this work. However, in order to judge the suitability of restoration models, historical maps (pre-dam period) were corroborated.

9.4 Result

Figure 9.2 displays the restorable sites in reference to the existing wetland sites. Both the models showed restorable sites are located within the major wetland tract of the river basin. Most of the cases the restorable sites are very contiguous to the existing wetland. However, the spatial extent of restorable sites is considerably higher in the case of MLC-based model (122 km^2). Restorable area was accounted about 41 km^2 in the case of SVM-based model. In this model, some very small patches were also appeared as like existing situation but its ecological and economic profitability could be evaluated before restoration process.

In the previous chapters, it has been established that dam-driven flow reduction is one of the cases behind dwindling flow in river, flood frequency, magnitude, and lateral extent. But this issue was not digitized at spatial scale. But if ecological flow is re-instated, flow volume is enhanced; these restorable sites will get more hydrological security and ecological connectivity leading to improvement in ecological service support.

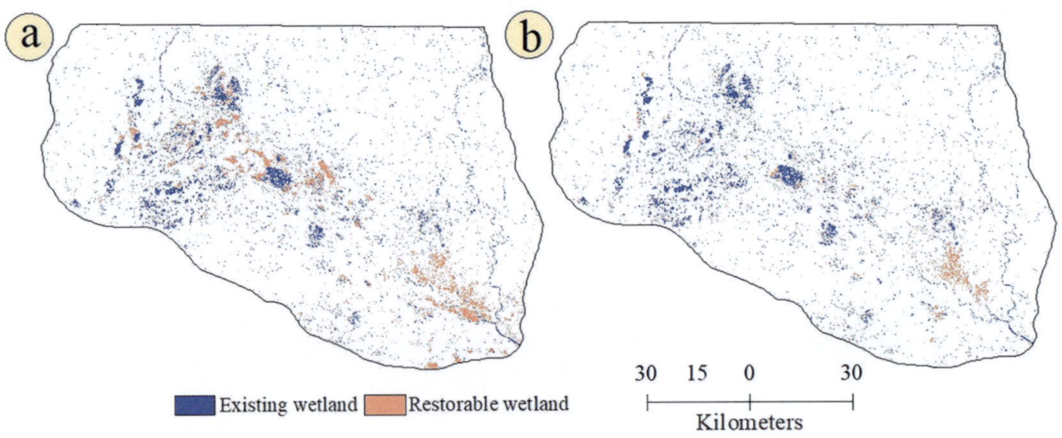

Fig. 9.2 Wetland restorable sites predicted using **a** MLC and **b** SVM

Other measures as mentioned below would also be helpful toward supporting restoration process.

9.5 Possible Ways Forwards for Wetland Management

(1) **River flow management**

Since the study clearly reported dam-induced flow lowering in Atreyee River has declined dam downstream flow lowering, flood frequency, magnitude, flood spilling extent, flow management would be a good step for restoring and surviving riparian wetland. In order to meet the need of growing water footprint, damming is one of the essential measures to enhance supply of water and energy. But the present study along with other existing related literature clarified that reduction in the flow may invite a series of short- and long-term penalties, including decrease in the carrying capacity, increase in downstream deposition, shrinkage in the active floodplain, de-linking and isolation of floodplain wetlands, disruption in the downstream aquatic ecosystem, growing eco-hydrological stress in wetland, drop in the base levels of the confluence streams, undue trapping of water in the head of the reservoir, squeezing channels, sporadic channel incision in some parts of the downstream reach, straightening of channels, and so on.

Taking all these adverse impacts in view, while damming upstream, it should be considered that downstream riparian processes, economies depend on river, so, significant depletion of water may deteriorate the river-dependent environmental processes, lives, and livelihoods. So, before damming, ecological flow should be computed and that flow should be allowed downstream for the survival of downstream river and riparian ecology.

Several scholars (Saha 1935 and Majumdar 1945) opined in opposite of constructing dam across rivers. In the quest of why did a river dry out?, Nobel laureate poet R. N. Tagore (Tagore 1917) condemned the attempts of arresting and restricting flow of rivers through dam and remarked it with the root incident that causing the death of a river in the subtropical region. Witnessing the enthusiasm and madness of damming and taming rivers across the world, he rightly remarked not to arrest a river or to use all its water, rather to keep the river flowing ever. In Chap. 3, flow exceedance of natural and different levels of flow modifications have been computed, so while decision making regarding how much water should be allowed downstream, this finding could be treated as crucial. Enhancing flow would increase flow availability in downstream river and hydrological connectivity between river and wetland.

(2) **Building spatial connectivity**

Study also revealed that due to reduction of flood spilling area, a good number of wetland is located beyond active flood spill area and they are not receiving regular flood water which is one of the main sources of water. Besides this, consistency of flood is also reduced significantly as discussed in Chap. 2. Due to this situation, floodplain wetlands are not getting abundant water from river. Reduction of water level is also caused reduction of seepage water connectivity between river and wetland. Lowering of groundwater level even below bed level of wetland distorted the vertical hydrological connectivity. In this circumstance, building connectivity between river and wetland (tie channel) and neighboring potential wetland could support water. Since the flow reduction is one of the major causes behind this, without eco-flow maintenance, it will not be fulfilled.

The perforation of agricultural land, built-up areas, and other forms of infrastructure associated with development has fragmented larger wetland patches into smaller wetlands. This mechanism itself has exposed such patches into human exposures. Spatial connections among these through at least some linear corridors could enhance the habitat quality. Larger patch is habitably potential than smaller patch (Xu et al. 2024). Even from the present study, it has been found that hydrological strength and eco-hydrological condition are substantially good in larger wetlands. So, maintenance of larger patches is eco-hydrologically relevant. So, minor modification in landscaping could enhance the eco-hydrological efficiency of wetland.

The United Nations World Water Development Report (2018) has acknowledged wetlands as an effective nature-based alternative to conventional gray infrastructure such as dams and reservoirs in watershed management frameworks. Restoring and enhancing the hydrological connectivity of wetlands has emerged as a cost-effective and sustainable strategy to address multiple ecological and hydrological challenges. Strengthening such connectivity is particularly beneficial in mitigating the impacts of extreme weather events—including erratic rainfall, floods, and droughts—and also aids in managing urban and agricultural effluents that degrade water quality. Improved hydrological linkages support the formation of aquatic systems that closely resemble natural conditions. These linkages play a pivotal role in maintaining biodiversity by influencing key ecological functions such as migration, reproduction, and dispersal of aquatic and semi-aquatic organisms (Schofield et al. 2018). Propagules—such as seeds, spores, and eggs of various taxa—are more efficiently distributed in systems where hydrological flows are uninterrupted. Thus, improving hydrological connectivity is essential for enhancing the ecological integrity and functional stability of wetlands. In this context, restoration of damaged or disrupted hydrological pathways becomes central to wetland rehabilitation efforts. A multifaceted restoration strategy is required, which may include restructuring river channels, reviving floodplain habitats, rebalancing sediment dynamics, ensuring adequate flow regimes, and improving water quality parameters. These interventions not only re-establish ecological processes but also contribute to the long-term sustainability of wetland ecosystems. A summary of potential restoration strategies, adapted from Cui et al. (2016), is presented in Table 9.1. It outlines key restoration types, their focus areas, and recommended engineering or ecological measures:

Implementation of these measures regionally and ecologically sensitive manner could significantly contribute to restoring wetland functionality. As a natural solution, they align well with sustainable development goals while ensuring long-term ecosystem resilience.

Table 9.1 Summary of wetland hydrological connectivity restoration strategies (adapted from Cui et al. 2016)

Restoration type	Key actions	Suggested measures
Channel morphology	Recreate natural channel and floodplain conditions	Removal of artificial reinforcements, slope regrading, riparian zone reclamation, buffer zone establishment, tie-channel reactivation
Lateral connectivity	Reconnect floodplain wetlands	Removal or modification of embankments, reduction in floodplain elevation, restoration of inundated habitats, re-vegetation
Longitudinal connectivity	Ensure uninterrupted river flows	Elimination of physical barriers like weirs and sills, construction of fish-friendly bypasses and widened bridge passages
Sediment dynamics	Re-establish sediment source-sink balance	Demolition or redesign of flow control structures, floodplain opening for sediment exchange
Flow regulation	Maintain ecological water levels	Supplemental water input from multiple sources to revive wetland hydrology
Water quality	Improve downstream water quality	Controlled water release, sediment flushing from reservoirs, and meeting ecological flow thresholds

(3) **Land use management**

The study region is agriculturally dominated. Since the floodplain is fertile and that could minimize production cost, people propensity is very high to cultivate shallower wetland and neighboring area. Hydrological security due to having wetland is another issue. In the study area, field investigation and image study clarified that majority of the wetland was reclaimed by agricultural land. In the present case, about 60% wetland area was reclaimed and existing wetlands are agriculturally dominated seasonally. These wetlands are some good hearth of summer paddy. Apart from that, seasonal grazing is another issue. Seasonal drying out of wetland bed and intensive grazing disturb the fixed soil carbon and thereby enhances soil erosion and carbon emission. Although this issue was not studied here, existing literatures on this (Limpert et al. 2021; Liu et al. 2022) proved this. Gradual deposition may cause for lowering water depth in deeper wetland bed. Paddy and jute cultivation in this region conventionally going on since long dates back. Intensively people try to do this. Aquaculture is an alternative economic option. In some parts of wetland in study example, lotus, makhana, fishes, etc. are some good provisioning products of wetland. MEA (2005) clearly reported that wetland could provide 2.5 times more monetary return in terms of its provisions. But lack of proper know how regarding feasibility of wetland use considering its ecosystem process often lead people toward reclamation of wetland. So two approaches could be taken including (1) restriction of wetland reclamation and (2) capacity building toward improving wetland-based cultivation.

(4) **Landscaping with native species**

One of the growing eco-hydrological problems is aggravation of fertilizer residues to the wetland and growth of water hyacinth. Exuberant water hyacinth growth hampers native species growth. In changing hydrological situation, strategies regarding modification in water hyacinth growth should be adopted. Accretion of agriculture residual within floodplain wetland is very difficult. Restriction of agricultural activities up to a certain distance (buffer zone) away from wetland could also partially reduce such accretion. Such zone could itself retain such pollutant to some extent. Maintenance of inlet and outlet system is also a good step in this way. Removal of hyacinth species and inspire native floral and faunal species who can survive in relatively lesser depth in modified hydrological condition would be a credential step toward managing ecological condition of wetland. It could partially help to restore ecological health of wetland. Different forms of native wetland species including emergent, floating, and submerged vegetations play significant role toward managing ecological ambience (Table 9.2). The existence of indigenous plant species exhibiting rich compositional variety, high biodiversity, well-developed structural form, and active ecosystem interactions serves as a strong indicator for assessing the ecological integrity and functional status of wetland systems (Kruse and Groninger 2003; Wilkins et al. 2003; Mitsch et al. 2012). Native vegetation not only reflects the ecological maturity of the site but also supports essential biogeochemical and hydrological processes that sustain wetland health. A wide range of aquatic plant species contributes distinctively to the functionality of wetlands. These species can broadly be grouped into emergent, submerged, and floating

Table 9.2 Ecological roles of common aquatic plant groups

Plant category	Representative species	Major ecological functions	References
Emergent plants	Cattail, reed, canna, calamus, fleur-de-lis, *Scirpustabernaemontani*	– Decomposition and nutrient uptake from water – Oxygen supply and microbial colonization via rhizosphere – Root absorption of heavy metals and pollutant retention	Sesin et al. (2021)
Submerged plants	Tape grass, hydrilla, *Cabombacaroliniana*, hornwort	– Nutrient absorption and control of eutrophication- Provision of habitat and refuge for aquatic fauna – Enhancement of dissolved oxygen levels	Eid et al. (2021)
Floating plants	Duckweed, *Nymphoides*, *Eichhornia crassipes* (water hyacinth), water lily	– Uptake of minerals from the water column – Promotion of algal growth – Rapid surface coverage to stabilize aquatic environments	Qin et al. (2021)

categories, each playing unique ecological roles. A brief summary of their ecological functions and representative species is provided in Table 9.2, adapted from Cai et al. (2021):

(5) **Linking ecosystem services with livelihood**

People's ignorance regarding wetland and often treating it as wasteland is a major inherent cause leading to loss and degradation of wetland. In this present case, hydrological modification was found crucial for wetland loss and eco-hydrological deterioration but people's knowledge about the potential ecosystem services is not sufficient even how to harvest these are also very lacking. If peoples get suitable livelihood option from wetland, people could try to save and restore wetland (Adhya and Banerjee 2022). Keeping wetland without any human intervention is very difficult when population density and demand of people are very high. So, protection of wetland absolutely in natural form may not be possible in this case. However, linking wetland ecosystem services with their livelihood is very pertinent step. This step could save the existing wetland and people's well-being.

9.6 Conclusion

Wetland's priceless ecosystem services toward natural system resilience and humanity could not be ignored anyway. With its wide range of immense ecological services, it could support USs Sustainable Development Goals (SDGs) particularly 2nd goal connoting zero hunger, 6th goal connoting clean water and sanitation, and 13th goal connoting climate action. Hence, conservation of wetland should be given priority. Literatures and present study unfortunately reported wetland loss and degradation of different forms of wetland health. It was further inferred that among different forms of wetland transformation and their linkages, hydrological transformation was very vital. Large-scale river flow modification along with other changes in hydrological conditions was reported very crucial. So, in order to conserve and restore floodplain wetland, eco-decision should be adopted and implemented. Ecological flow should be continued downstream of dam. This is very crucial. This flow would help to minimize other forms of wetland transformation rates. Ample water availability in wetland could restrict people for easy reclamation of wetland. Along with this, improving wetland literacy (proper knowledge about wetland, its functions, benefits, burdens) and capacity building for sustainable wetland use are also very vital steps for protecting wetland. Planners, regional, and local stakeholders should be come forward in order to make this project successful. Since the river and wetland alongside are cross border (India and Bangladesh), bilateral cooperation is also necessary.

Wetland conservation is a global imperative. The World Wetlands day 2025 revolved around the theme "Protecting wetlands for our common future" prioritizing the interconnectedness of wetland conservation and human well-being. The main goal of damming and hydrological modification of river is also catering humanity. Wetland itself caters natural system and human being at larger extent. In order to fish some benefits from damming spending huge money, we compel to forego a wide array of ecosystem services from natural river flow as well as dependent wetlands. Moreover, for getting small and short-term benefit from damming, the society is going to deprive from long term and wide range of priceless benefits. Decision makers, even common people have to rethink about such conflicting decision and act judiciously for establishing environmental justice. Scientific research and environmental and societal impact assessment are very urgent before any dam construction.

In essence, it could be treated that dam acts as a profound determinant of river flow and riparian ecology shaping hydrological, ecological landscape, rhythms of flow, and ecosystem pulses. The influence of it promotes hydro-ecological risk to wetland in short and long-term manner. So, an ecological agreement between river flow and wetland in floodplain area is very urgent. As wetlands offer crucial benefits and support wetland-dependent livelihoods, it is urgent to assess the ecological footprint of damming, ensuring that such interventions promote wetland recovery and resilience.

References

Adhya T, Banerjee S (2022) Impact of wetland development and degradation on the livelihoods of wetland-dependent communities: a case study from the lower gangetic floodplains. Wetlands 42(7):65

Ausseil AG, Gerbeaux P, Chadderton WL, Stephens T, Brown D, Leathwick J (2008) Wetland ecosystems of national importance for biodiversity: criteria, methods and candidate list of nationally important inland wetlands. Landcare Research Contract Report LC0708/158

Cai S, Song X, Hu R, Guo D (2021) Ecosystem-dependent responses of vegetation coverage on the Tibetan Plateau to climate factors and their lag periods. ISPRS Int J Geo-Inf 10(6):394

Cui H, Shi J, Qiu L, Zhao Y, Wei Z, Wang X et al (2016) Characterization of chromophoric dissolved organic matter and relationships among PARAFAC components and water quality parameters in Heilongjiang, China. Environ Sci Pollut Res 23:10058–10071

Dai M, Su J, Zhao Y, Hofmann EE, Cao Z, Cai WJ et al (2022) Carbon fluxes in the coastal ocean: synthesis, boundary processes, and future trends. Annu Rev Earth Planet Sci 50(1):593–626

Eid EM, Shaltout KH, Al-Sodany YM, Haroun SA., Galal TM, Ayed H, et al (2021) Temporal potential of Phragmites australis as a phytoremediator to remove Ni and Pb from water and sediment in Lake Burullus, Egypt. Bull Environ Contam Toxicol 106(3):516–527

Johnston CA, McIntyre NE (2019) Effects of cropland encroachment on prairie pothole wetlands: numbers, density, size, shape, and structural connectivity. Landscape Ecol 34:827–841

Karim F, Dutta D, Marvanek S, Petheram C, Ticehurst C, Lerat J et al (2015) Assessing the impacts of climate change and dams on floodplain inundation and wetland connectivity in the wet–dry tropics of northern Australia. J Hydrol 522:80–94

Karim F, Marvanek S, Merrin LE, Nielsen D, Hughes J, Stratford D, Pollino C (2020) Modelling flood-induced wetland connectivity and impacts of climate change and dam. Water 12(5):1278

Kruse BS, Groninger JW (2003) Vegetative characteristics of recently reforested bottomlands in the lower Cache River Watershed, Illinois, USA. Restor Ecol 11(3):273–280

Limpert KE, Carnell PE, Macreadie PI (2021) Managing agricultural grazing to enhance the carbon sequestration capacity of freshwater wetlands. Wetlands Ecol Manage 29:231–244

Liu Z, Deng Z, Davis SJ, Giron C, Ciais P (2022) Monitoring global carbon emissions in 2021. Nat Rev Earth Environ 3(4):217–219

Majumdar SC (1945) Rivers of Bengal delta, Department of Irrigation, Government of West Bengal, pp 45–98

Mao D, Wang Z, Wu J, Wu B, Zeng Y, Song K et al (2018) China's wetlands loss to urban expansion. Land Degrad Develop 29(8):2644–2657

MEa, M. E. A. (2005) Ecosystems and human well-being: wetlands and water synthesis

Meng XW, Li Y, Shen L, Yang XQ (2020) Reinforcing a water bridge in a disjoint nanochannel. Europhys Lett 131(2):20003

Mitsch WJ, Bernal B, Hernandez ME (2015) Ecosystem services of wetlands. Int J Biodivers Sci Ecosyst Serv Manag 11(1):1–4

Mitsch WJ, Wu X, Nairn RW, Weihe PE, Wang N, Deal R, Boucher CE (1998) Creating and restoring wetlands. Bioscience 48:1019–1030

Mitsch WJ, Zhang L, Stefanik KC, Nahlik AM, Anderson CJ, Bernal B et al (2012) Creating wetlands: primary succession, water quality changes, and self-design over 15 years. Bioscience 62(3):237–250

Odoh BI, Nwokeabia CN (2024) Impact of land use and land cover changes on groundwater dynamics in selected local government areas of Anambra State, Nigeria. Int J Earth Sci Knowl Appl 6(2):131–142

Peng Z, Yu X, Li S (2024) Aggradation and reworking of an alluvial fan in response to climate changes on the south bank of Lake Qinghai, NE Tibetan Plateau. J Asian Earth Sci 264:106073

Qin S, Zhang X, He S, Huang J (2021) Improvement of nitrogen removal with iron scraps in floating treatment wetlands. Environ Sci Pollut Res Int 28(14):17878–17890

Rodrigo MA (2021) Wetland restoration with hydrophytes: a review. Plants 10(6):1035

Saha MN (1935) Collected works of M.N. Saha, University of Calcutta

Schofield O, Brown M, Kohut J, Nardelli S, Saba G, Waite N, Ducklow H (2018) Changes in the upper ocean mixed layer and phytoplankton productivity along the West Antarctic Peninsula. Philos Trans Royal Soc A Math Phys Eng Sci 376(2122):20170173

Sesin V, Davy CM, Freeland JR (2021) Review of Typha spp.(cattails) as toxicity test species for the risk assessment of environmental contaminants on emergent macrophytes. Environ Pollut 284:117105

Sievers M, Hale R, Morrongiello JR (2017) Do trout respond to riparian change? A meta-analysis with implications for restoration and management. Freshw Biol 62(3):445–457

Spieles DJ (2022) Wetland construction, restoration, and integration: a comparative review. Land 11(4):554. https://doi.org/10.3390/land11040554

Tagore RN (1917) Durakshanka poem of novel Citra, Visva-Bharati

Tomscha SA, Bentley S, Platzer E, Jackson B, de Roiste M, Hartley S et al (2021) Multiple methods confirm wetland restoration improves ecosystem services. Ecosyst People 17(1):25–40

United Nations (2018) World Water Development Report 2018: nature-based solutions for water. UNESCO,

Paris. https://unesdoc.unesco.org/ark:/48223/pf0000261424
United States Environmental Protection Agency (US EPA) (2007) Integrated risk information system. US EPA
Wilkins S, Keith DA, Adam P (2003) Measuring success: evaluating the restoration of a grassy eucalypt woodland on the Cumberland Plain, Sydney, Australia. Restoration Ecol 11(4):489–503
William W (1930) Lectures on the ancient system of irrigation in Bengal, University of Calcutta
Xu J, Xu B, Xia GS, Dong L et al (2024) Patched line segment learning for vector road mapping. In: Proceedings of the AAAI conference on artificial intelligence, vol 38, No. 6, pp 6288–6296
Yang Z, Cai Y, Mitsch WJ (2015) Ecological and hydrological responses to changing environmental conditions in China's river basins. Ecol Eng 76:1–6
Zedler JB (2000) Progress in wetland restoration ecology. Trends Ecol Evol 15(10):402–407

Index

S. Pal et al., *Damming and Wetland Eco-hydrology*, GIScience and Geo-environmental Modelling, https://doi.org/10.1007/978-3-032-11729-8

MIX
Papier aus verantwortungsvollen Quellen
Paper from responsible sources
FSC® C105338
FSC
www.fsc.org

If you have any concerns about our products,
you can contact us on
ProductSafety@springernature.com

In case Publisher is established outside the EU,
the EU authorized representative is:
Springer Nature Customer Service Center GmbH
Europaplatz 3, 69115 Heidelberg, Germany

Printed by Libri Plureos GmbH
in Hamburg, Germany